RATIONAL ACCIDENTS

Inside Technology

Wiebe Bijker and Rebecca Slayton, series editors

A list of books in the series appears at the back of the book.

RATIONAL ACCIDENTS

RECKONING WITH CATASTROPHIC TECHNOLOGIES

JOHN DOWNER

The MIT Press
Cambridge, Massachusetts
London, England

The MIT Press would like to thank the anonymous peer reviewers who provided comments on drafts of this book. The generous work of academic experts is essential for establishing the authority and quality of our publications. We acknowledge with gratitude the contributions of these otherwise uncredited readers.

This book was set in Stone Sans and Stone Serif by Westchester Publishing Services. Printed and bound in the United States of America.

Library of Congress Cataloging-in-Publication Data

Names: Downer, John (John R.), author.
Title: Rational accidents : reckoning with catastrophic technologies / John Downer.
Description: Cambridge, Massachusetts : The MIT Press, [2023] | Series: Inside
    technology | Includes bibliographical references and index.
Identifiers: LCCN 2023002845 (print) | LCCN 2023002846 (ebook) | ISBN
    9780262546997 (paperback) | ISBN 9780262377027 (epub) |
    ISBN 9780262377010 (pdf)
Subjects: LCSH: Reliability (Engineering) | Aircraft accidents—Prevention. |
    Risk assessment. | Industrial accidents—Prevention.
Classification: LCC TA169 .D69 2023 (print) | LCC TA169 (ebook) |
    DDC 620/.00452—dc23/eng/20230202
LC record available at https://lccn.loc.gov/2023002845
LC ebook record available at https://lccn.loc.gov/2023002846

Dedicated to Dick, Babs, Greenie, and the Loris.
My family.

CONTENTS

ACKNOWLEDGMENTS

I spent far too long writing this book. What began as an exciting research project ended as a sort of weird, life-consuming mental illness. I have excuses. Changing jobs, each with new obligations, things like that. I did write a lot of other stuff along the way. A ruptured brain aneurism didn't help; traumatic brain injuries are bad for productivity in my experience. (Although I'd recommend them to anyone looking to invisibly raise the difficulty settings on life.) But probably a bigger issue than all that were the book's outsized ambitions. I had an elaborate map for its argument worked out fairly early on, and in retrospect that argument had too many moving parts. I use myself as a cautionary tale for my graduate students; the next book will follow my own advice to them and have a much simpler structure. It's also true that delay bred delay—the later this became, the better I felt it had to be to justify the lateness. It took me a minute to realize I was always going to be hopelessly underwater on that ledger.

Which is all to just to say that I've been at this a long time and accrued a lot of people I need to thank.

The biggest regret, at least in regard to lateness, is that the book's two biggest academic influences and advocates—Trevor Pinch and Charles Perrow—both passed before I was able to thank them here. I owe them each an enormous debt. I moved to Ithaca to study under Trevor. He became the lead supervisor of my PhD, which, in turn, became the seed from which this book grew. And it was he who suggested that I write about the seemingly paradoxical epistemology of civil aviation safety. Perennially cheerful, idiosyncratic, and supportive, Trevor routinely helped chivvy the drafting of this manuscript along over the years; offering insightful comments on very early drafts of most of its chapters. It was a great loss to STS in general, and to me personally, when he died so prematurely. Charles Perrow—or "Chick" as he preferred—became a mentor, friend, and (for a time) colleague, after someone sent him a paper I'd written for anonymous review. He somehow figured out who wrote it so he could reach out to talk to me. I was an admirer of Chick's work, and the gesture meant a lot to me at an early point in my career. He helped me arrange a move to California the following year, where I had the pleasure of getting to know him better. With his enormous curiosity and indomitable generosity of spirit, Chick did academia right. He must have commented on dozens of my drafts over the years—something he did for a lot of people, I think—and always with eagerness, warmth, humility, and insight. He had an outsized influence on this volume, the title of which is an homage to his most relevant and best-known work. I only met him in the twilight of his life, but I miss him, and sociology is poorer without him. He and Trevor were good men.

Very honorable mentions also go to a few others. Sveta Milyaeva, in particular, helped drag this book over the finish line. She read and commented on a late draft of every chapter at least once, sometimes twice, and became the manuscript's chief advocate and cheerleader. It's a huge debt I owe her. Hugs also to her partner, Nestor, who gamely said he'd be interested in reading a draft and then actually did—which never happens in my experience. Kudos to his stamina there. I should also thank Carl Macrae, my fellow nerd, who must have read a lot of different elements of this over the years. He's the busiest man I know, but on the occasions that I can pin him down on something work-related, he always has insightful ideas and encouraging words. And thanks as well to Dr. Adnan Siddiqui—an extremely nice neurosurgeon in upstate New York, who fixed my uninsured brain with two rounds of surgery and an expensive crocodile clip. He was a little skeptical that I'd ever be fit for book-writing, but I promised I'd give him a nod in my acknowledgments. I really appreciate the help, everyone.

Those are the highlights, but there are many other people I want to recognize here. I'll try to err on the side of inclusivity, so it'll run a bit long. But I don't think anyone reads acknowledgments unless they're looking for their own name so I doubt it matters. Just skip ahead if you're bored. (And if I've forgotten anyone, please know that I'll feel bad about it.) For the sake of structure, I'll organize you all chronologically, starting with the book's roots in graduate school.

Top of my Cornell thanks should go to Michael Lynch and Ron Kline, who were on my dissertation committee with Trevor. Like Trevor, both were, and I'm sure remain, brilliant scholars and profoundly good people. I was lucky to work with them. In retrospect, there were several other people at Cornell STS who I regret not getting to know better while I was there. Steve Hilgartner's writing, in particular, has been a big influence on me in recent years. Sonja Schmid and Anna Mareker made up the rest of my diminutive yeargroup; I'm glad they're doing well. Other STS colleagues included Cyrus, Jamie, Hans, Carin, and Dave, who were all fun to know. More broadly, my thanks and best wishes to all my Stewart Little housemates (especially Guillermo Mendoza and Pete Buston, plus Patrick, Soazig, Kevin P., Jacques, Nicole, Gadi, Dave, Megan, Eric, Erica, Zach, Rochelle, James, and Aaron). And, beyond SLC, to Kerry Papps, plus Ross, Lauren, Erin, Laurent, Andrei, Kelly, Kevin O., Valerie, Tamar, Malou, Morten, Andy, Julie, Matt, Kathy, Danielle, Jenka, and Amanda. Thanks also to Terry Drinkard, who was a big supporter of the dissertation over this period and provided a lot of useful feedback and insight. I hope he's doing well.

Then I was in London with a hole in my head. I wasn't at King's College London's Centre for Risk Management for long, but I need to thank Henry Rothstein: another good man and fine scholar, who took a gamble in hiring me as his postdoc. I wasn't a great flunky—I was pretty sleep deprived most of the time, and I jumped ship before the project ended, which must have been exasperating—but Henry was very understanding. I owe him a debt. Sebastian and Kristian also were good friends and colleagues there. Moving onto LSE's Centre for Analysis of Risk and Regulation (CARR). Bridget Hutter hired me and always had my back in her own unique way. She was another supporter of the book who sadly passed before it was finished. I think we underestimated, at the time, how much she did to protect CARR's staff and its research culture. My colleagues Carl, David, Anna, and Jeanette all became good friends. And, outside the academy, Paul Yates was the cornerstone of my social life. London was a weird but memorable time.

On leaving London there was Stanford's Center for International Security and Cooperation (CISAC). I was probably out of place as a security researcher, but these were remarkable years for which I will always be grateful. I learned a lot of unexpected things about nuclear weapons and the US security establishment in general, some of them absurd and most of them pretty terrifying. I got to walk the site of the first atomic explosion, which was memorable, and I was almost, kind of, vaguely spied on by the Russians, which still amuses me. Fukushima happened while I was there, starting a chain reaction that ended up derailing this book for a long time (although I think the argument is stronger for it). Lynn Eden, another scholar I admire, was the heart and soul of the fellows program there; I can't imagine it's the same without her. Scott Sagan unknowingly cost me a case of cheap champagne by not being drafted into government. My fondest CISAC memories are of the other fellows, especially Matthias Englert, lately owner of cheap champagne, and Anne Harrington, who probably still doubts America's appetite for electric vehicles. Beyond them I also want to thank Ryan and Erin, Rob and Bekah, and Jan. Plus Toshi, Gaurav, David, Pablo, Ed, Jason, Katja, and Brenda. Also Rebecca, who made comments on a very early chapter of this manuscript, long before she accidentally became one of its editors. Outside of CISAC, Ross Halvorson was generosity itself; I will always be grateful for his support and friendship. I should also thank Peter Ladkin—whom I happened to meet in person during this period—for originally posing, in an email to Trevor Pinch, the question that inspired this book. He has the keenest mind for engineering epistemology of anyone I know (even if he probably wouldn't call it that).

This brings us to Bristol's School of Sociology, Politics, and International Studies (SPAIS), where I've now been for longer than I'd ever anticipated. The new public management mores of mainstream British academia came as a shock when I joined, and I remain an odd fit in a department that's far more interested in social justice than technological risk (and more power to them). SPAIS tolerates me with grace, however, and Bristol is a neat city to call home. I'm lucky to have worked alongside some great colleagues here. Special mention goes to Miriam Snellgrove and Jess Paddock: fellow "losers," along with Sveta. I would never have survived the pandemic without them. I'm glad to have Tom Osborne occasionally manage my line. In his broad erudition and gently disheveled, well . . . everything, he evokes a simpler time in British academia when nobody used phrases like "pathway to impact." Special mention also to Paul and Nivi, Maud and Ryerson, Jeremy, Junko, Jess O., Tim E., Beckie, Therese, Claire, Max, Alexis, Paul and Pete. Thanks also to all the staff of Coffee #1 on Welsh Back, who let me take up space, year after year, while

I tinkered at this manuscript. I've met a lot of people outside Bristol during my time here as well; too many to mention. I will, however, single out Diane Vaughan, who helped set up a rewarding stint in New York (my self-inflicted visa issues notwithstanding). And, especially, Ramana, whom I'll always admire for so successfully navigating academic bureaucracy without the benefit of a surname (also for his remarkable scholarship, productivity, patience, friendship, and many other fine qualities—but mostly the surname thing).

Returning to the book itself, I should thank my incredibly patient editors of the Inside Technology Series at the MIT Press. Also, my considerable thanks to three anonymous reviewers, whose identities have been fun to guess (five, if you count the series editors). All gave generous, insightful, and useful feedback on the original manuscript, which is significantly stronger for their input. None of you liked the original title; I hope the new one feels like a better fit. Titles are hard.

This just leaves my extended family. A man couldn't ask for nicer step-relatives than Jen, Jan, and Steve. Nor could he ask for more accommodating in-laws than Richard and Alicia, or Krissy and Kevin (and the kids), who have always made me welcome in their homes. I wish that Bristol could be closer to Lake Superior. Will, Rose, and Jack—my perfect nephews and niece—are far too young and sensible to have any interest in this book, but I hope they might one day get a tiny kick from having been mentioned in it. Pomps the cat is an adorable ball of fur and exasperating worrier of ankles; he brings me peace. Dave is sadly missing, but never forgotten.

Last, and the very opposite of least, my parents, brother, and wife, to whom this book is collectively dedicated. I'm so deeply grateful for everything they have done for me that nothing I could say here would ever feel remotely adequate. So I'm just going to say that I love you all and let the dedication stand alone. For, as Wittgenstein put it, "whereof one cannot speak, thereof one must remain silent."

INTRODUCTION: PURPOSE, SCOPE, AND STRUCTURE

We live in a society exquisitely dependent on science and technology, in which hardly anyone knows anything about science and technology.
—Carl Sagan

PURPOSE

What's wrong with knowing what you know now and not knowing what you don't know now until later? It's a meditative question, artfully phrased by A. A. Milne for Winnie the Pooh. But the answer—at least if you're designing a complex technological system with the potential to fail catastrophically—is "everything." If somewhere, deep in the bowels of your sprawling atomic weapons arsenal and infrastructure, there lurks an unknown software bug (or an unanticipated corrosion process, or anything else) capable of accidentally instigating a thermonuclear war, that's something you need to know now. Because finding it out later, the hard way, could lead to you having what Milne would probably have called "a bit of a day."

This should be a sobering thought, even for stoics like Pooh. With ever more confidence and enthusiasm we have been building technologies that—like our nuclear weapons infrastructures, and for essentially the same reasons—need to be known with great certainty *now* rather than *later*. These are technologies that we absolutely cannot allow to fail catastrophically, lest we incur intolerable hazards. The only way to be certain that such technologies will not fail, however, is to deeply understand every aspect of their functioning, and there are very convincing epistemological reasons to believe that this depth of understanding is fundamentally unachievable. The tools that experts would use to interrogate them—the tests, models and calculations by which we know machines—are all inescapably imperfect, and it is rational to imagine that this imperfection should, very occasionally but still too frequently, give rise to unavoidable catastrophes. If knowing the limits of our knowledge is itself a form of wisdom, in other words, then there is a strong argument to be made that we are pursuing some of our technological ambitions in folly: trusting our lives and livelihoods to an implausible promise of technological reliability.[1]

The significance of this conclusion is obvious, but it immediately raises a difficult question because experts demonstrably *have* mastered extreme reliability in at least one highly complex technology: jetliners. In the civil aviation sphere, if in no other, they appear to have transcended the imperfections of their tools, and the limits of their knowledge, by building intricate and demanding machines with failure rates that are as low as we require of our most hazardous systems. We can know this with confidence because, unlike most technologies that require extreme reliability, we operate jetliners in high enough volume to measure their failure rates statistically. And although they do still fail, very infrequently, these failures are diluted by their numbers; such that equivalent failure-rates would imply near perfect reliability in systems like reactors, which we operate in much smaller numbers.

This "aviation paradox," as I will call it, is the problem that drives the argument that follows. The extraordinary reliability of jetliners, together with the accurate predictions of that reliability made by aviation regulators, speaks to an astonishing depth and breadth of understanding of the myriad intricacies of jetliner design and operation. An understanding that is so deep and so broad, I will argue, that it ought to be incommensurable with what philosophers have told us about the limits of engineering knowledge. Civil aviation's achievements in this regard raise fascinating epistemological questions, therefore, and the answers to the questions have important implications regarding our relationship to hazardous technologies more broadly. As we will see, jetliners are not reliable for the reasons we are led to believe, and the way experts achieved this reliability is less transferable than we imagine.

In exploring this question, the chapters that follow introduce a novel way of thinking about technological accidents. They make the case that some accidents—which I will call "rational accidents" in this book (but have elsewhere referred to as "epistemic accidents" [Downer 2011b; 2020])—are most usefully be attributed to the fundamental limits of what engineers can know about the technologies they build. Such accidents might not be common, I will argue, but they have several noteworthy properties. Much like Perrow's (1999 [1984]) well-known "normal" or "system" accidents, but for meaningfully different reasons, they cannot be avoided, even with ideal organizational practices and engineering analyses. And insofar as they can be considered "blameworthy" at all, it is only in the sense that experts should have understood the limits of their control. Unlike normal accidents, however, they can be managed over time: incrementally reduced, almost (but never entirely) to the point of being eliminated. But only if the organizations managing them have access to certain resources and are willing and able to follow certain practices, the costs of which can be prohibitive.

The argument explores the implications of these rational accidents. It outlines their significance and logical necessity, it examines the ways in which experts have learned to manage them in civil aviation, and it considers the transferability of their achievement to other technological domains. More generally, however, it explores the epistemological problems posed by extreme reliability in complex systems. And with these goals mind, there are a few ways in which its scope is perhaps broader, and simultaneously narrower, than might immediately be apparent. Before proceeding further, therefore, two points are worth clarifying in this regard: one regarding the argument's theoretical ambitions, the other its practical relevance.

SCOPE

THEORETICAL AMBITIONS The argument that follows is probably narrower with regard to its theoretical ambitions than might be intuitive. In light of this, it is useful to briefly clarify its relationship to two distinct academic literatures: the Science and Technology Studies (STS) literature, and the wider social-scientific literature around technological safety and disaster.

The text has a slightly unorthodox relationship to the STS literature. It is firmly grounded in the STS tradition in the sense that it builds on the epistemological limitations of seemingly objective knowledge and explores the subjectivities that arise from those limitations. Where a more conventional STS account might invoke those subjectivities to historicize and contextualize the knowledge claims it is discussing, however, the emphasis here is more on the practical difficulties and dangers to which they lead. The intention, in essence, is to explore how (and if) expert organizations successfully navigate the subjectivities of extreme technological reliability; it is not to explain how they came to understand and/or manage extreme reliability in the manner that they do. The latter questions are undoubtedly worthwhile, and their answers make for fascinating stories. The structures and practices through which modern societies govern the reliability of extremely hazardous technologies are more contingent, contested, and socially constructed than we commonly realize (see, e.g., Jones-Imhotep 2017; Wellock 2021; Johnson 2002). They are just not the stories being told here. The STS scholarship that this book draws on most directly, and hopefully most closely resembles (at least in spirit), are texts like Mackenzie (1990; 1996a, b; 2001), Wynne (1988), Collins (1988), Pinch (1991; 1993), and Collins and Pinch (1998); all of which, in different ways, explore the indeterminacies of engineering knowledge and its social implications.

The text has a more complex relationship to the social scientific literature around technological safety and disaster. In exploring the epistemology of extreme reliability, it does speak directly to some core concerns of this literature—the question of what makes technologies safe, for instance, and of why they sometimes fail—but its ambitions in this regard are nevertheless limited. By this, I mean that I am not attempting to grapple with every aspect of what makes technologies with catastrophic potential safe.

Decades of social research into this question has shown that safety is always a holistic accomplishment requiring constant work on a wide range of fronts: as much a function of organization and culture as of engineering epistemology. And it has shown that technological accidents have equally layered and complex underpinnings. Silbey (2009) offers a useful overview of this literature, but Vaughan's (2021) exploration of air traffic controllers, which shines a light on some of the nuanced practices underpinning safe air travel, is exemplary and pertinent (see also, e.g., Perrow 1999; Reason 2016; Dekker 2011; Schulman 1993; Perin 2005; Snook 2000; Vaughan 1996; Turner 1978; Sagan 1993; Weick and Sutcliffe 2001). Seen from the perspective of this literature, the decision to examine technological safety and disaster through the prism of engineering epistemology, as this volume does, might appear reductive.

It is important to understand, therefore, that the account that follows very intentionally explores only a narrow, albeit vital, dimension of technological safety: the question of how it is even *possible* for experts to know complex systems well enough to achieve and verify of them very extreme reliabilities. The book unpacks this question; it looks at how engineers have managed such tasks in a specific domain, and it considers the wider implications of their success. That is all. It is still a lot, however, in the sense that the narrowness of the central question does not preclude some strong conclusions with far-reaching implications. Ultrareliable design will never be a *sufficient* condition of technological safety, but in the context of technologies that absolutely cannot be allowed to fail it is a *necessary* condition. And while failures attributable to the limits of engineering knowledge cannot sufficiently explain all accidents, they can explain some accidents. (Note that safety and disaster are asymmetrical, in that the former requires everything to go right, while the latter requires only that one thing go fatally wrong.) The many organizational and cultural dimensions of building, maintaining, and operating complex technological systems might be essential to their safety, therefore, but if we cannot also master the epistemological questions then they will never be enough.

Even if the question being explored here represents only one piece of the safety problem, therefore, it is nevertheless a vital piece, at least in certain contexts. And on this basis, I will speak often about the safety of some technologies being "dependent" on specific conditions and practices that allow experts to push beyond the limitations of their tests and models. Where I do so, however, the intention is never to imply that those practices or conditions are *sufficient* for achieving safety. Researchers seeking a more rounded and expansive understanding of how technological safety is accomplished will need to read more widely.

With that caveat, therefore, this volume might reasonably be counted amid the wider social-scientific scholarship on technological safety and disaster— enough so, at least, that it is probably instructive to locate it in relation to that literature. Such positioning is invariably fraught; not least because it is difficult to map the academic discourse on any complex topic without distorting that discourse. Scholars have nuanced understandings of their work's distinctiveness, and understandably chafe at being lumped together into rough-hewn categories. Navigation requires some systemization, however, and even the most flawed maps can be useful. So, at the risk of reifying some contested categories, I think it is enlightening to understand the argument in relation to an old but influential schema that divides the academic safety literature along an axis that speaks to some of its core themes.

This schema originated with Sagan (1993), who proposed that social scientific scholarship on technological safety might usefully be divided into two broad schools, based, in essence, on whether they hold that technologies can ever be made perfectly safe. On one side, Sagan (1993: 13) posited an "optimistic" school, which construed safety as fundamentally perfectible. This school is exemplified, in his telling, by what he calls the Berkeley group (e.g., La Porte and Consolini 1991; Roberts 1989; Rochlin et al. 1987; Schulman 1993; Rochlin et al. 1987), but it also encompasses outliers like Wildavsky (1988), and today would almost certainly include works such as Vaughan (2021), as well as the burgeoning (largely European) "safety science" literature (e.g., Le Coze 2020). On the other side, he posited a "pessimistic" school, which construed safety as fundamentally imperfectable (Sagan 1993: 13). This school is exemplified in his account by Perrow (1999 [1988]), but it also includes texts such as Clarke (1989), Shivastava (1987), and Reason (1990), and today would probably also include works like Vaughan (1996), Snook (2000), as well as Sagan (1993) itself.[2] In the form that Sagan's schema is conventionally remembered (and largely invoked by Sagan himself), however, the different schools are usually narrowed to just his exemplars—the Berkeley group and Perrow—whose arguments are cast as opposing theories of disaster: Berkeley's optimistic High-Reliability Theory (HRT) being set against Perrow's pessimistic Normal Accident Theory (NAT).

Sagan's schema is overreductive, to be sure, especially in its narrowed form. It ascribes an unwarranted degree of coherence to heterogeneous bodies of literature, and a lot of research fits imperfectly into his categories. (It is notable that many of the researchers he describes rejected his interpretation of their position on perfectibility; most notably his optimists [e.g., LaPorte 1994; LaPorte and Rochlin 1994]). For all its limitations, however, his construal of the literature was undeniably influential. In part because his book—compellingly written, with eye-opening accounts of close calls with accidental atomic war—was revelatory and persuasive. But also, and more substantially, because his categories did seem to capture a real, tangible divide in the scholarship itself; a divide that still has relevance today.

To appreciate the lasting value in Sagan's schema, it helps to think of his schools as divided less by the question of perfectibility, which Sagan foregrounds, and more by whether they are primarily oriented towards researching safety or failure. On one side, we might say, is scholarship that takes, as its primary objective, the question of what makes complex systems safe, and seeks to identify the characteristics of safe systems. This is Sagan's optimistic literature, exemplified in his account by HRT. And on the other side there is scholarship that takes, as its primary objective, the question of why complex systems fail, and seeks to identify the characteristics of unsafe systems. This is Sagan's pessimistic literature, exemplified by Perrow and his normal accidents.

These two projects heavily overlap, of course. "Safety" in this context can almost be defined as the absence of failure, so the difference between researching one and researching the other is often a matter of perspective. Perspectives are important, however, and it is fair to say that the differently-oriented research questions tend to shine meaningfully different lights on the same phenomena. They encourage dissimilar questions, assumptions and interpretations, and, as a result, they tend to yield different insights, opinions and conclusions. So it is, for example, that scholars exploring safety—Sagan's optimists—tend to see the potential for organizations to ameliorate human frailties, counteract economic pressures, and learn from experience. Whereas those exploring failure—Sagan's pessimists—are more likely to see the potential for organizations to be undermined by human frailties, perverse incentives, and imperfect learning. Where one side might portray a "glass of safety" as "99 percent full," as Sagan (1993: 48) puts it, the other is more likely to portray it as "1 percent empty."

This difference in perspective, when extended to Sagan's core question of the perfectibility of safety, gives rise to contrasting pictures, even if it can be difficult to fix points of fundamental disagreement. (When pushed, both sides tend to agree that organizations are capable of developing impressive stratagems for safety that are always imperfect at the margins, and that both the impressiveness and imperfections are worthy of scrutiny.) As Sagan intuited, the distinction becomes most acute in the context of Perrow (1999 [1988]) specifically. This is because Perrow, almost uniquely, makes a sustained and principled argument about the inherent unavoidability of certain kinds of accidents, and then explores the implications of this inevitability. (These are his "normal accidents," the logic of which chapter 7 will explain in more detail.) It is also because Perrow, like Sagan himself, focuses much of his analysis on a technology with such extreme failure hazards that even a 1 percent safety deficit takes on enormous significance (reactors in Perrow's case, and atomic weapons in Sagan's). It is still difficult to find points of fundamental disagreement. Perrow is not claiming that all, or even most, accidents are unavoidable, and safety-oriented scholars are not explicitly claiming the opposite. At the same time, however, it is rare for the latter to highlight or unpack the implausibility of perfect safety, or to grapple with its implications for extremely hazardous technologies.[3] As with the engineering discourse around such systems, most safety scholarship frames technological failure as a *problem to be solved* rather than an *inevitability to be confronted.*[4]

Sagan's portrayal of the literature offers an instructive backdrop against which to understand the argument of this book, not because it easily accommodates that argument, but because it doesn't. The argument that follows addresses many of the core issues he identifies—the evitability of accidents; the perfectibility of extreme safety; the potential for organizational learning—but in ways that cut squarely across his basic divide.

It would be easy to pigeonhole this volume as belonging squarely in Sagan's disaster-oriented, pessimist, tradition. The title is an allusion to Perrow (1999 [1988]), after all, and there are several clear parallels with that text. It is primarily focused on technologies with catastrophic potential, for instance, wherein even tiny shortfalls in the "glass of safety" become extremely important. And the opening two parts (seven chapters) examine why experts inevitably struggle to fill that glass. As with Perrow's text, moreover, this examination cumulates in the proposition of a new type of accident—different from but analogous to his "normal accident"—that is fundamentally unavoidable, in the sense that even perfect organizations (if they existed) would have no way of preventing it. More broadly, the argument also reckons with the limits of organizational rationality: emphasizing the power of incentive structures to shape practices and priorities, even at the expense of safety.

For all this, however, a case could also be made that much of the text belongs in the safety-oriented, optimist, side of Sagan's ledger. This is because the argument effectively switches orientation in its second half, moving from examining the causes of failure to exploring the achievement of safety. This is to say that, having explored the epistemological dilemmas of extreme reliability, it then turns to explore how civil aviation manages to transcend these limits. It argues that the accidents to which those limits give rise are only fundamentally unavoidable in certain circumstances, and explains how those accidents—if tolerated and interrogated over long periods—can serve as a foundation on which to build extreme reliability. Then, in chapters that might almost be considered High Reliability Theory, it unpacks the specific practices and structures that, with appropriate incentives, enable organizations to leverage that foundation.

Perhaps unsurprisingly, therefore, the conclusion sets a slightly ambiguous tone, especially with regard to the question of how optimistic we should be about technological safety. In the vast majority of technological contexts, I see no reason for undue pessimism on this front. As with Perrow, I am not suggesting that all (or even most) technological disasters are unavoidable, or that organizational safety practices could not be honed to reduce the frequency of such disasters. And, to the extent that some failures are unavoidable, I do not believe that these failures are necessarily very significant in most contexts. Outside a narrow range of technologies with catastrophic failure potential, failures do not always imply disasters, and it is often reasonable to believe that Sagan's glass can be made sufficiently full, even if it can never be filled entirely.

When it comes to technologies that never can be allowed to fail, however, I will argue that we must be more cautious. The levels of reliability that we require of such technologies are actually achievable, given the right conditions and practices. Jetliners prove this. But civil aviation is exceptional in regard to its conditions and practices. Outside of that specific domain, we are not managing our most failure-intolerant systems in ways that would allow the reliability we demand of them. And, in almost all cases, we would not be able to do so. The way that we conventionally understand the reliability of jetliners leads us to think that same reliability should be achievable elsewhere, but a fuller understanding of this achievement implies the opposite conclusion.

PRACTICAL RELEVANCE   If the argument is perhaps narrower in its theoretical ambitions than might be intuitive, then it is also probably broader in its applications and implications. Much of its focus is on jetliners and civil aviation, but this is only because they offer a unique window into a wider epistemological problem regarding extreme reliability in complex systems more broadly. It is in this broader context—of what the safety of modern air travel implies for other complex systems with catastrophic potential—that the argument has real purchase. The practices and logics through which we manage ultrareliable technologies closely resemble each other, and all must grapple with the same epistemological dilemmas, so exploring one illuminates the others.

This is not to say that the argument has no bearing on more mundane technologies with less demanding reliability requirements. This is especially true of the idea at its center, the rational accident, which in principle might apply in almost any technological context. Not every accident is rooted in the fundamental limits of engineering knowledge, of course, and most are better understood through a different lens. But rational accidents are not especially uncommon—as we will see, for instance, they are almost certainly more prevalent than Perrow's normal accidents—and I can imagine the idea potentially being useful in a range of situations.

There are several reasons why, in this volume, I have chosen to explore the idea exclusively in the context of extremely hazardous technologies. One is simply that technologies that cannot be allowed to fail offer an ideal lens through which to explore the mechanisms and implications of inevitable failure. If some accidents are inherently unavoidable—as this argument claims—then this fact has obvious purchase in circumstances where accidents must be avoided at all costs.

Another is that rational accidents are easier to identify and substantiate in this context. Experts scrutinize catastrophically hazardous technologies extremely closely and subject their performance to elaborate assessment and oversight. So on occasions when design shortcomings cause these technologies to fail, it is more difficult to attribute those failures to insufficient effort, rigor or organization, and it becomes more credible to attribute those failures to the fundamental limits of what effort, rigor and organization can achieve. (As a general rule, we might suppose that the proportion of failures best attributed to epistemological limitations rises as technological systems become more reliable overall.)

A third reason for framing the argument in this way is simply that the primary objective of this volume was always to explore the epistemological problem of extreme reliability, more than it was to explore the causes of accidents. Rational accidents are at the heart of its argument and are probably its most generalizable insight, but the intrinsically important problem of extreme reliability is its organizing theme. Insofar as others find the idea of rational accidents useful, therefore, I leave it to them to explore its more diffuse implications and its applications in wider technological contexts. Books need to end somewhere.

STRUCTURE

In keeping with its various turns, the argument that follows has a slightly unusual structure. The heart of the text grapples with the work of making jet-liners reliable, but the narrative opens and closes with discussions of nuclear reactors, particularly the 2011 Fukushima meltdowns. This "nuclear sand-wich," as one reviewer called it—with reactors as the bread and jetliners as the bacon, lettuce, and tomato—might almost seem like a bait and switch, but it serves a useful purpose. Most directly, it underlines the point that the argument's real significance lies outside of civil aviation. Nuclear reactors are not the only other complex technology from which we demand extreme reliability, as we will see, but they make an excellent counterpoint to jetlin-ers. More clearly than any other system, they exemplify why civil aviation's reliability achievements matter to our lives and livelihoods, and illustrate why the nature of those achievements should be a cause for concern about other technological domains. If understanding jetliners offers us insight into extreme reliability, we might say, then understanding reactors gives that insight meaning.

Between its nuclear bookends (or bread slices), the argument is organized methodically, as a series of subarguments that build on each other progres-sively. Some of these elements could stand alone, but the wider argument does not lend itself especially well to readers who would skip around. With this in mind, I have labored to make the structure of the wider argument as transparent as possible. To this end, for instance, the chapters that follow are organized into parts. The opening chapters, grouped in part I, outline the core conceptual problem and establish its significance: they make a prin-cipled argument that ultrahigh reliability should not be possible in complex technologies, and then problematize this claim by establishing that jetlin-ers are, in fact, ultrareliable. The chapters in part II unpack that problem: they look at the processes by which jetliners are ostensibly made reliable, illustrating the epistemological limitations of those processes and invoking those limitations to articulate the idea of unavoidable "rational accidents." Those in part III resolve that problem: they look behind civil aviation's ostensible reliability processes to explain how experts actually manage the epistemology of jetliner reliability in practice. And those in part IV explore the wider implications of that resolution: they examine the transferability of civil aviation's reliability achievement and reflect on what it means for our relationship to other technical systems with catastrophic potential.

Such structuring might seem heavy-handed at times, but it will hopefully help readers keep the argument's larger trajectory in mind even as the narrative itself sometimes navigates more winding roads. For those who might become lost along the way, however, I will close this introduction with a more detailed chapter breakdown to which they might refer.

CHAPTER BREAKDOWN

PART I

- *Chapter 1* invokes the 2011 meltdown at Fukushima to introduce the idea of "catastrophic technologies": complex, sociotechnical systems requiring ultrahigh reliability (i.e., mean-times-to-failure in the region of billions of hours of operation). It argues that the proliferation of these technologies has made the need for experts to achieve, and predictively verify, such extreme levels of reliability a consequential and underappreciated dimension of modern governance.

- *Chapter 2* draws on epistemology from STS and the philosophy of science to argue that the extreme reliability required of catastrophic technologies should be impossible for experts to achieve and to verify. It proposes that the processes that experts use to know these technologies contain too many indeterminacies—too many qualitative judgments— to support useful predictions of a complex system's failure behavior over billions of hours of operation to a satisfactory degree of certainty.

- *Chapter 3* problematizes the argument of chapter 2 by finding that experts do, demonstrably, achieve and accurately verify ultrahigh levels of reliability in jetliners. It calls this contradiction the "aviation paradox," and argues that it makes jetliner reliability practices uniquely interesting.

PART II

- *Chapter 4* outlines the structures, logics, and practices through which US regulators ostensibly govern the extreme reliability of civil jetliners. It finds that the indeterminacies of these practices are visible as ambiguities in the reliability requirements to which jetliners are held.

- *Chapter 5* uses engine bird-strike testing as a case study through which to illustrate the inherent limitations of testing for reliability more broadly. It highlights a range of uncertainties about the representativeness of bird-strike tests, and shows how these uncertainties give rise to doubts about their meaning. It argues that all technological tests necessarily grapple with the same dilemma, and that this should preclude assertions of ultrahigh reliability.

- *Chapter 6* follows a similar pattern as chapter 5, but in relation to theoretical models (as opposed to empirical tests) of a jetliner's reliability. It uses redundancy calculations as a case study through which to illustrate the inherent uncertainties of reliability calculations and the doubts to which they give rise. Tests and models might serve distinct purposes at the regulatory level, it argues, but they are subject to the same underlying epistemological constraints.

- *Chapter 7* is the heart of the argument in many ways. It draws on the inevitable indeterminacies of tests and models, outlined previously, to explain why some catastrophic failures—what it calls rational accidents— will necessarily and unavoidably elude even the most rigorous engineering analysis and oversight. It then compares this argument to that of normal accident theory, which similarly says that some accidents are unavoidable. Both perspectives imply inevitable failures, it concludes, and neither precludes the other, but the mechanisms (and thus the implications) of each are meaningfully distinct.

PART III

- *Chapter 8* resolves the paradox of why jetliners can be so reliable despite the indeterminacies of engineering knowledge and the rational accidents that they imply. The key to understanding jetliner reliability, it argues, is to recognize that civil aviation does not manage reliability in the manner it purports. Rather than using formal assessment practices—tests and models—to examine new machines, it finds that experts have slowly whittled their understanding of jetliners over time, assiduously interrogating failures to hone a common, stable, airframe design-paradigm to extreme levels of reliability.

- *Chapter 9* builds on chapter 8 by further substantiating its most controversial claim: that the reliability of modern jetliners depends on them adhering to a common and stable design paradigm. To this end, it looks first at new composite materials, which are often considered evidence of meaningful innovation. It contextualizes this technology and the opinions held about it, arguing that composites represent a more modest and incremental shift than is immediately apparent, and that opinions about it need to understood in context. From there, it turns to consider instances where airframers have unambiguously embraced radical innovation. It looks at military aviation, arguing that although military aircraft are highly innovative, their reliability is substantially inferior to that of civil jetliners as a result. Finally, it addresses Concorde. It recognizes that the airplane was, without doubt, a radically innovative design, but finds that this was reflected in its counterintuitively dismal reliability record.

- *Chapter 10* introduces an organizational problem that arises from the solution to the epistemological aviation paradox. It observes that the stability of airframe designs implies that the companies that design jetliners often forgo, or substantially delay, adopting economically advantageous innovations. And that, in doing so, they consistently trade short-term competitive advantage for long-term safety: behavior that organizational sociologists have long held to be unrealistic. Having outlined and substantiated this new problem, the chapter goes on to resolve it. It looks at regulation but concludes that the ambiguities of technological practice make this an implausible explanation. It then revisits and reassesses civil aviation's structural incentives, arguing that certain characteristics of the industry (primarily its operating volume) give it a unique relationship to failure that incentivizes reliability to a much greater degree than in other catastrophic technological spheres.

- *Chapter 11* substantiates and refines the argument of chapter 10, concerning the primacy of structural incentives. It first addresses the 737-MAX crisis. The MAX, it argues, amply illustrates the extreme costs of unreliability in civil aviation—and thus the industry's unusual incentive structure—but it also reminds us that organizations are complex and imperfect, and that (absent periodic shocks) they tend to lose sight of their interests. From there, the chapter turns to consider civil aviation's relationship to crash survivability. It finds that the industry is markedly less proactive about design choices intended to make jetliner accidents less dangerous than it is about those intended to make accidents less frequent. This difference exemplifies the primacy of structural incentives, it claims, as the incentive structures around crash frequency (or reliability) and crash survivability pull in opposing directions.

PART IV

- *Chapter 12* begins to explore the wider implications of civil aviation's approach to managing extreme reliability. It proposes that the public portrayal of jetliner reliability management—as grounded in formal analysis via tests and models—is misleading and elides the real practices and conditions on which ultrahigh reliability depends. It then explores the reasons and logic driving this misportrayal, finding that it sometimes can be functional in civil aviation. From there, however, it argues that the misportrayal also gives rise to underappreciated costs and dilemmas, many of which are likely to be less acute in civil aviation than they are in other catastrophic technological domains.

- *Chapter 13* focuses on one prominent problem that arises from the misportrayal of civil aviation's reliability practices: the false but widespread belief that its achievements in this regard should be replicable in other catastrophic technological domains. Understood properly, it argues, the ultrahigh reliability of jetliners depends on resources and practices—such as high operating volumes, a commitment to design stability, and a legacy of instructive failures—that are not available, practicable, or achievable elsewhere. The most consequential cost of misportraying jetliner reliability management, it concludes, is that this difference becomes opaque.

- *Chapter 14*, by way of coda, returns to Fukushima. It invokes the accident to argue that reactors exemplify the essential differences between jetliners and other catastrophic technologies. It argues that experts working in the civil nuclear sphere enjoy few of the resources that their counterparts in civil aviation use to navigate the harsh epistemology of extreme reliability. Although ostensibly governed via equivalent structures and processes to jetliners, it concludes, reactors cannot be as reliable. This difference is difficult to observe because of the small number of reactors in operation, it argues, but the shortcomings of reactors are nevertheless visible if we know where to look—not least in Fukushima itself, which might reasonably be understood as a rational accident.

1 THE AVIATION PARADOX

Wherein it is argued:

- That in some technological domains it has become critical that experts achieve and predictively verify extreme levels of reliability **(chapter 1)**
- That these levels of reliability should be impossible for experts to achieve and verify in such systems **(chapter 2)**
- That experts demonstrably achieve and verify these levels of reliability in civil jetliners **(chapter 3)**

1 CATASTROPHIC TECHNOLOGIES: THE RISE OF RELIABILITY AS A VARIABLE OF CONSEQUENCE

... by slight ligaments are we bound to prosperity and ruin.
—Mary Shelley

1.1 INTRODUCTION

A NEAR-MISS

From halfway around the world, and with the passage of time, the 2011 Fukushima Daiichi nuclear disaster can appear almost unremarkable—just one more entry in the sad ledger of technological failures. This is understandable. The plant's three reactor meltdowns were dramatic while they lasted, certainly, but the world did not stop in 2011, and even in most areas of Japan, daily life rebounded relatively quickly. But this apparent normalcy is deceptive. Understood properly, Fukushima—as I will henceforth refer to the accident—was a near-incomprehensible catastrophe. It should have been a wake-up call to a world that has stumbled blindly into an increasingly dangerous relationship with its technological creations.

Appreciating Fukushima's full significance is difficult, as many of its ramifications remain opaque. Estimates of its costs—human, financial, and ecological—have all swollen considerably in the years since 2011,[1] but in slow and incremental steps that have rendered the rising numbers and their implications all but invisible (Downer 2016).[2] Revised pollution estimates and reassessed expenses rarely sell newspapers, especially when those estimates are unusually complex or contested like Fukushima's. Media coverage of the disaster has been a disorienting crossfire of claims and counterclaims—a product of the esoteric issues involved and the powerful interests at stake (see, e.g., Hamblin 2007). Even relatively uncontested claims can be rhetorically impotent in this context; the numbers involved are either too large to be relatable—millions and billions occupying a similar space in the collective consciousness, (a phenomenon that psychologists sometimes refer to as "scope neglect" [e.g., Kahneman 2000])—or too abstract to be meaningful. (No measure of radiological pollution makes for an intuitive headline, whether expressed in "rads," "rems," "roentgens," "becquerels," "grays," "sieverts," "curies" or even "bananas.")[3]

To cut through this complexity and contestation around Fukushima, some commenters try to convey the accident's true gravity by focusing on a single, authoritative fact with intuitive significance. For Naoto Kan, the Japanese prime minister at the time of the crisis, that fact was that Japan came bracingly close to losing its capital city.

Proponents of atomic energy robustly dismiss claims that Fukushima threatened Tokyo as hyperbole. Japan's capital is about 150 miles (241 kilometers) from the site of the accident, and few experts contend that the meltdowns could ever have released enough pollution to jeopardize its residents. However, in Kan's telling, and that of many other experts, the meltdowns were not the main event.

Kan's account of how the city narrowly escaped catastrophe centers instead on one of the plant's spent-fuel pools. The pool in question, one of several at the site, was housed above an inoperative reactor in unit 4 of the plant. Resembling a 45-foot (14-meter)–deep swimming pool, its purpose was to store radioactive materials safely. At the time of the accident, it contained years' worth of new and spent reactor fuel: 1,535 assemblies, each consisting between fifty and seventy individual rods. These rods had to be kept submerged in cold water lest they combust spontaneously. Under normal circumstances, this would not have been challenging. Four days into the crisis, however, a large explosion in the reactor building severely damaged the pool and its supporting structures. At this point, the pool began to leak, and keeping the rods submerged in water became extremely challenging.

Plant workers battled desperately to refill the water faster than it drained, but the circumstances were punishing. Desperately understaffed because most of their colleagues had fled the site's soaring radiation, and woefully underequipped, they were simultaneously managing three concurrent meltdowns amid the destruction and disruption of a historic tsunami and earthquake.

The stakes, however, could not have been higher. Had the pool's fuel rods been exposed for long, they would have ignited and burned unquenchably, liberating an unprecedented volume of radionuclides into the atmosphere. (Among other radioactive materials, the pool contained ten times the volume of cesium-137 released by the Chernobyl accident in the mid-1980s.) This would have forced a full evacuation of the plant, which in turn would have led to the loss of its other pools, some of which were considerably larger. The hazards of this "devil's chain reaction," as one commenter would later refer to it, would have been far more extreme than those of the meltdowns alone. Extreme enough to have very plausibly necessitated the evacuation of Tokyo City (Gilligan 2016; Sieg and Kubota 2012; Lochbaum, Lyman, and Stranahan 2014, 80–85; Osnos 2011; Dvorak 2012a, 2012b; Matsumura 2012; Goodspeed 2012; Fackler 2012; Lean 2012; Kan 2017).

Industry advocates vigorously dispute this conclusion, but it is supported in substance by an extensive range of highly credible sources. It is now clear, for instance, that the US Nuclear Regulatory Commission (NRC) believed such a scenario to be possible, and at points even probable. Indeed, its chairman later testified to Congress that he believed it had come to pass. An interagency team was convened to plan the emergency evacuation of US citizens from Japan's capital (Lochbaum et al. 2014; Osnos 2011). The official, independent report on the disaster prepared for the National Diet of Japan (NAIIC 2012) is clear about the danger. Its release was accompanied by a flood of chilling testimony. "We barely avoided the worst-case scenario, though the public didn't know it at the time," the chairman of the inquiry told reporters (Lean 2012). "It was extreme luck that Japan managed to avoid experiencing the most disastrous day" said another prominent member (in Lean 2012).

If "extreme luck" seems like perverse commentary on a triple reactor meltdown amid a natural disaster of historic severity, then consider the implications of Japan's unrealized "disastrous day." The gravity of Japan losing Tokyo is difficult to capture in a language that cheerfully wastes its superlatives on breakfast cereals and laundry detergents, but it would undoubtedly be immense. Cities are surprisingly resilient, but radiological pollution is insidious and difficult to clear. Japan's capital, which once rose phoenixlike from the ashes of horrific US fire bombings of World War II, might have been felled much more permanently by its government's own energy policy; Fukushima's fallout forever poisoning the earth, like the salt that Rome plowed into the soils of Carthage.

Even a temporary evacuation—which the Japanese government began planning in earnest (Gilligan 2016)—would have taken a momentous human and economic toll. Tokyo is a gigantic city, home to over thirty-five million people (more than New York and London combined). Like those cities, it is a command center of the global economy, and it has the highest gross domestic product (GDP) of any metropolis on the planet. In a world where even mispriced mortgage derivatives can instigate a financial meltdown and "Great Recession," the potential financial fallout of such an event almost defies comprehension. At the very least, it would have jeopardized Japan's ability to service its national debt. Kan might not have been exaggerating much when he told the *Wall Street Journal* that his country's "existence as a sovereign nation was at stake" (Quintana 2012).

So it is that Fukushima—already remembered as the signal technological disaster of recent decades—is arguably best understood as a near-miss. Had the explosion in unit 4 been even slightly larger, had individual plant workers been less courageous, or had the area been hit by a significant after-tremor (which are common in the wake of earthquakes), then the pool would likely have been lost, with all the consequences that loss implies. Experts can debate the likelihood of this scenario, but probable or not, the fact that it was even *possible* should cast a long shadow over humanity's relationship to technical ambition in the twenty-first century.

BETTING ON TECHNOLOGY

There are many lessons to be drawn from Fukushima, but perhaps the most fundamental of them pertains to the authority afforded to engineering safety assessments and the purveyors of those assessments. Almost unknowingly, Japan bet its future on the understanding that engineers could speak definitively about extraordinary failure probabilities in a complex technological system, and it almost lost everything.

At some point in our recent history, nations started building technologies that simply could not be allowed to fail catastrophically—technologies with the latent capacity to acutely imperil lives, livelihoods, and lifestyles on a massive, unconscionable scale. It is important that we govern such technologies wisely. This, in turn, requires that we understand those technologies. Or, perhaps more meaningfully, it requires that we understand the nature of our relationship to them: the capabilities and limitations of the expert bodies that we charge with overseeing them, along with the provenance and credibility of the safety assurances that those bodies provide.

These capabilities and assurances, as well as their fundamental limitations, are the central theme of this book.

1.2 CATASTROPHIC TECHNOLOGIES

A NEW RELATIONSHIP TO FAILURE

Catastrophic technological failure has been a meaningful public concern for as long as there have been engineered structures. In December 1879, outside Dundee in Scotland, the Tay Rail Bridge collapsed during an evening storm, dropping a train full of passengers into the icy waters of the Firth of Forth. In March 1864, outside Sheffield in England, the newly constructed Dale Dyke Dam burst as its reservoir was being filled; the ensuing flood damaged over 600 houses and killed over 240 people. In January 1919, a huge molasses storage tank in Boston erupted at its seams, unleashing a sugary, 35-mile-per-hour tidal wave that fatally engulfed 21 people. There are many such stories, some dating back many hundreds of years. The Koran speaks of the failure of the "Ma'rib dam" in the year 570 or 575, which, historians believe, could have displaced upward of 50,000 people from what is now Yemen.

Despite this history, however, there remains a sense in which modernity's relationship to technological failure changed meaningfully in the mid-twentieth century. As it is only at this juncture that advanced industrial societies began to build complex, dynamic sociotechnical systems—as opposed to static structures—that could almost never be allowed to fail.

I propose to call such systems "catastrophic technologies."

Let us define catastrophic technologies as complex technological systems that—because of the potential hazards (mortal, social, economic, or environmental) of them failing catastrophically (i.e., in a way that leads to a major accident)—would not be viable (economically or politically) unless the probability of such failures was deemed low enough to justify excluding them from all public discourse and decision-making. They are complex systems that require extraordinary, and historically unprecedented, failure rates—of the order of hundreds of millions, or even billions, of operational hours between catastrophic failures.

This performance requirement is the defining feature of catastrophic technologies—even more than the extreme failure-hazards from which that requirement stems. In most cases, it must be known in advance of these technologies being allowed to operate. At minimum, it must be assumed as a premise of their operation: implicit in discourse and decision-making that never even consider catastrophic failure as a possibility. Every debate and decision regarding catastrophic technologies is grounded in implicit or explicit assumptions about failure being, for all intents and purposes, a solved problem. States would not tolerate such technologies unless they considered catastrophic failures to be functionally impossible, or close enough to impossible as to be beneath serious policy consideration.[4]

Technologies requiring such extreme failure behavior arose as the product of two interrelated trends. The first, straightforwardly, was the emergence of complex systems with the potential to fail in ways that were so unprecedentedly hazardous or costly that even a single, isolated failure might lead to intolerable harm. Reactors exemplify this phenomenon, as Fukushima amply illustrates. The second, slightly more complicated, trend was a sharp rise in the adoption of complex systems that could tolerably be allowed to fail (unlike reactors), but only so long as those failures remained extremely infrequent. Jetliners are exemplary here. Jetliner crashes are tragic, no doubt, but their isolated costs are not equivalent to those of reactor meltdowns. So in the early days of civil aviation—when infrequent flights limited the absolute number of accidents, and smaller aircraft limited the absolute number of people who could die in each accident—the socially tolerable number of accidents per departure could be (astonishingly) high by modern standards. At this time, airliners did not require the extreme mean time to failure of a catastrophic technology. As jetliners emerged and air travel became much more democratic, however, the relative frequency of accidents had to fall precipitously for air travel to remain viable.[5] There were 51 fatal commercial airplane accidents in 1929, roughly one for every million miles flown. This was deemed acceptable at the time, but the same rate today would imply over 7,000 fatal accidents per year: just short of 20 every day (and with many more passengers on each flight). As the absolute number of flights rose, therefore, the failure requirements on airliners became ever more demanding, to the point where they emerged as catastrophic technologies.

This is all to say that that, over roughly the same period (from the 1950s to the 1970s), reactors and jetliners both came to require extraordinary levels of failure performance—making them catastrophic technologies—but for slightly different reasons. Reactor meltdowns were (and remain) far more consequential than jetliner crashes, so they needed to be commensurately less probable. No state would tolerate a Fukushima-scale accident on its soil in exchange for atomic energy, but all would accept a tiny number of jetliner crashes in exchange for mass air travel. With far fewer reactors in operation than jetliners, however, both systems needed similar failure probabilities to satisfy these different requirements. The absolute number of plane crashes it would take to jeopardize the political viability of jetliners is clearly far higher than the absolute number of meltdowns it would take to jeopardize the political viability of reactors, in other words, but we require broadly equivalent levels of failure performance from both technologies because there are many more jetliners than there are reactors.[6]

(In light of the distinction made here, and for reasons that will become apparent as the argument progresses, it is useful to divide catastrophic technologies into two types on the basis of why they require their extreme failure performance. Let us call these types "chronic" and "acute." Where chronic catastrophic technologies, exemplified by jetliners, are those systems that can tolerably be allowed to fail on rare occasions but require their extreme failure performance because of the volume at which they operate. And acute catastrophic technologies, exemplified by reactors but representing the majority of all catastrophic technologies, are those that operate at much smaller volumes but require extreme failure performance because *any* failures are intolerable. Both chronic and acute have equivalent failure requirements, we might say, but meaningfully different relationships to individual failures.)

Catastrophic technologies are still quite rare, although this is changing. Reactors and jetliners were early and exemplary examples, but in the years since World War II, modern societies have become increasingly dependent on the near-infallible functioning of various complex technological systems (Perrow 2007). Certain industrial plants, laboratories, and drilling platforms require commensurately extreme failure performance, for instance, as do a significant number of slightly less sociotechnical systems, such as undersea communications cables and some medical devices. Atomic weapons pose exquisitely acute demands in this regard, albeit in a meaningfully different political and organizational setting that—in the context of this argument, at least—makes them less exemplary than reactors and jetliners.[7] And, insofar as we wish to construe networks as technologies in themselves, as we almost certainly should, then so do their sprawling command-and-control infrastructures, the failure of which could instigate atomic wars (and almost have on multiple occasions [Sagan 1993; Schlosser 2013]). Along the same lines, we might also recognize a range of other networks: from banking computers and electricity grids to air-traffic control and global positioning system (GPS) networks. The internet, in particular, increasingly represents both a catastrophic technology in itself and an acute new source of vulnerability in many other systems (Kaplan 2016). Broadening the definition of "technology" in a different direction (and stretching the terminology of "failure"), a wide range of nonmechanical technologies have similar catastrophic potential, and thus require equivalent assurances about their behavior: some novel biological agents (Downer 2020), for instance, and financial instruments like the derivatives that played a crucial role in the 2007–2008 financial crisis (MacKenzie 2005).

As with most social scientific categories, however, the term "catastrophic technology" defies precise enumeration. "Catastrophic-ness" is neither a binary nor a wholly objective condition, so the exact tally of qualifying technologies will always be contestable. The examples given in this book are in no way intended to represent a comprehensive list, therefore. It is reasonable to assume that others are woven invisibly into the fabric of modern life, and that their numbers will expand in the coming years. (Consider, for instance, autonomous vehicles [Stilgoe 2018], artificial intelligence systems [Bostrom 2014], or even Moon habitats [Benaroya 2018], all of which are under active development.) The picture is further muddied by domain-specific terminologies and preconceptions. (We rarely speak about biological or economic disasters in terms of "systems failure," for example.) For all the ambiguity and heterogeneity of catastrophic technology as a category, however, I would argue that the mutual need for extreme failure performance that defines it creates enough meaningful commonalities between different systems—in their governmentality, and (especially) their epistemology—for the idea to be useful.

Catastrophic technologies have common properties, therefore, but exploring those properties requires depth and detail, which in turn demand focus. For this reason, the argument that follows will focus on a narrow range of electromechanical systems: primarily jetliners and reactors, with most of the emphasis on the former. The logic for this was outlined in the preface and will become more evident as the argument progresses. Broadly, however, these two specific technologies have been chosen for three reasons. The first is that neither is an ambiguous or borderline example of a catastrophic technology; both exemplify the category and its essential qualities. The second is that the structures—organizations, laws, rules, practices, terminologies, norms—through which both are governed are similar enough to make comparisons intuitive. And the third is that, despite their ostensible similarities, the underlying constraints on each technology are distinct enough to make comparisons useful. They might be governed via ostensibly similar structures, but they are subject to very different epistemological limitations, and this fundamentally alters the way those structures must operate. So it is, I will argue, that contrasting the reliability practices around jetliners with those around reactors offers revelatory and widely generalizable insights into the governance of all catastrophic technologies.

Let us turn, then, to examine the structures through reactors and jetliners are governed.

1.3   GOVERNING CATASTROPHIC TECHNOLOGIES

THE PRIMACY OF RELIABILITY

In the US and beyond, the catastrophic technological era has given rise to a distinctive form of governmentality, the properties and priorities of which reflect the unique demands of extreme failure hazards. Most straightforwardly and intuitively, this governmentality is characterized by an unusual emphasis on risk management: risk being a far more prominent concern in the governance of jetliners and reactors than in that of toaster ovens and televisions. Distinctively, however, risk management in this context has come to be dominated by concerns about *failure* risks, which tend to eclipse risks associated with systems' normal operation. This is simply to note, for instance, that the safety of reactors or jetliners is conventionally understood almost exclusively in terms of the probability of them failing catastrophically, even though there are ways for them to be unsafe without failing (by polluting, for instance) (see Rijpma 1997; Wolf 2001).

The convention of construing risk exclusively in terms of failure is made more distinctive by a parallel convention of construing failure exclusively in terms of probability. In most engineering contexts, failure risks are understood as a function of both their probability and their consequences. (Such that engineers might improve the safety of systems by making failures less hazardous when they occur, as well as less likely to occur in the first place.) In catastrophic technological contexts, however, the potential hazards of failures are all but intolerable, so the probability of failure is king. (So it is, for instance, that we address the risk of accidental nuclear detonations by ensuring that they never occur, not by digging bunkers and relocating populations.) The potential consequences of failure are rarely forgotten entirely in this context—as we will see, for instance, states mandate crash survivability measures in jetliners—but such considerations are always secondary and attenuated. This is all to say that there is little talk of "resilience" in the discourse around catastrophic technologies, as the bureaucracies responsible for managing them place far greater weight on preventing failures than they do on mitigating their hazards. If technological artifacts can be understood as responses to implicit problems, as Baxandall (1985) suggests, then the key problem to which catastrophic technologies are a response is how to avoid accidents, not how to survive them.

In electromechanical contexts, these all-important failure probabilities are usually metricized as "reliability," defined narrowly here as the frequency of catastrophic failures (as opposed to, for instance, the frequency and duration of unplanned downtime), and usually expressed as a mean-time-to-failure. For instance, the viability of both jetliners and reactors hinges on them exhibiting a known mean-time-to-failure north of hundreds of millions of hours (tens of thousands of years), or what I will henceforth refer to as "ultra-high reliability."

Again, this convention of emphasizing reliability is more distinctive and less inevitable than might be intuitive, not least because electromechanical systems can fail catastrophically for reasons that would not traditionally be construed as "reliability": sabotage, for example, or human error. Such considerations are rarely forgotten entirely. Indeed, states assiduously manage issues like security or human performance in reactors and jetliners. Yet high-level discussions of these systems' safety are predominantly framed as engineering problems, wherein reliability measures are widely treated as expressions of the absolute likelihood of catastrophic failure from any cause. "Soft" difficult-to-measure risks, such as might arise from security or human-performance questions, are usually folded into these reliability metrics, either by treating them as solvable problems, which functionally disappear once certain requirements have been satisfied, or by treating their probabilities as objectively quantifiable, akin to those of mechanical failure.[8]

Here, then, are four noteworthy characteristics of the modern (arguably Western) approach to governing catastrophic technologies:

1. It places a strong emphasis on controlling their risks (more often referred to as "safety").

2. It predominantly construes their risks in terms of their failure behaviors.

3. It largely understands their failure behaviors in terms of probability alone (addressing the likelihood of failures rather than their consequences).

4. It tends to reduce the probability of them failing to a reliability metric.

Later chapters of this book will explore these characteristics in greater detail and examine some of their implications. For now, however, it suffices to recognize the significance of reliability as a metric by which modern states measure catastrophic technologies, and through which they approach the governance of those technologies. Directly or indirectly, almost all public deliberations about catastrophic technologies hinge on expert reliability calculations.

So it is that the catastrophic technological era has elevated technological reliability to a variable of enormous consequence. There is undoubtedly a degree of elision in this elevation. Reliability does not capture every dimension of failure, failure does not capture every source of risk, and risk is not the only meaningful measure of a catastrophic technology. But even if reliability cannot encompass everything—or even everything that is ascribed to it—the emphasis afforded to it is not wholly inappropriate. Safety is the sine qua non of catastrophic technologies, and reliability is a necessary component of that safety. Absent at least an implicit understanding of reliability, any other claim about such technologies becomes moot. Reactors and jetliners can only be "economical" or "environmental" to the degree that they do not explode, melt down, or fall repeatedly from the sky.

OBJECTIVE AND AUTHORITATIVE

For all its newfound prominence, the reliability of a complex system—especially at the ultrahigh levels required of catastrophic technologies—is far from being a transparent variable. One cannot simply examine the reliability of a system as one would its weight, volume, or length. Establishing reliabilities at such levels, making them visible and auditable, is an elaborate process, fraught with difficulties (as chapter 2 will explain in detail), and requiring extensive expertise. So it was that the emergence of reliability as a variable of political consequence was accompanied by the parallel emergence of expert intermediaries charged with making it publicly accountable. (Sims [1999] calls such bodies "marginal groups" to connote the fact that that they inhabit more than one social world, translating one to the other and performing the boundary work of differentiating credible from untrustworthy knowledge.)

Along with atomic energy and mass air travel, therefore, the mid-twentieth century saw the birth of elaborate reliability-accountancy structures: organizations, metrics, rules, guidance, and practices, all designed for measuring and ensuring ultrahigh reliability. The modern incarnations of these structures have many national and domain-specific idiosyncrasies, but most have substantial underlying commonalities. In the US, for instance, their contours are broadly equivalent across the civil aviation and civil nuclear domains and have close analogs in non-US contexts. Their most visible manifestations are dedicated design assessment and oversight bodies, usually prominent subdivisions of larger regulatory agencies, such as the Federal Aviation Administration (FAA), which oversees US civil aviation, and the NRC, which oversees US atomic energy.

These technology oversight bodies wield more power than is often apparent. As scholars of audit processes have long attested, experts and their calculative practices become highly influential when an important property is knowable only through its assessment (Hopwood and Miller 1994; Power 1997), and reliability is no exception in this context. The inherent inscrutability of extreme reliability, together with its centrality to catastrophic technological discourse, imbue the bodies that account for it with considerable agency, which is bolstered by the fact that audiences conventionally construe the findings of these bodies as being highly authoritative: the product of precise, objective, and rule-governed processes. Publics and policymakers routinely doubt and question the findings of expert economists or sociologists, but once the NRC or FAA officially measures the reliability of a specific reactor or jetliner, then a broad range of institutional actors, from courts to budgetary planning offices, are all but obliged to treat that measure as an established fact.

The convention of treating catastrophic technological reliability assessments as objective facts should not be surprising. It is firmly in keeping with a pervasive cosmology of technoscientific knowledge, wherein the work of interrogating machines is routinely assumed to be driven by strict methodologies, which, when performed diligently and correctly, yield knowably correct results (Rip 1985; Jasanoff 1986; Wynne 1988; Mitcham 1994; Ezrahi 2008). Modern bureaucracies, we might say, operate within a positivist cosmology wherein catastrophic technological black boxes have knowable properties, the truth of which can be established definitively via the checkboxes of formal audit practices.

This positivist cosmology, with its associated certainties, is far from unique. Indeed, it can be found wherever public policy intersects with the properties of technological artifacts. When applied to the reliability of catastrophic technologies, however, it becomes especially crucial. For here, more than anywhere, it is vital that expert determinations be construed as impersonal, authoritative, and rule-governed. Where the fates of cities are at stake, assertions of technological reliability need to stand on more than the considered opinion of industry insiders. On such questions, if on few others, certainty and objectivity are nonnegotiable.

It should be troubling, therefore, that in this specific context—where reliability claims are made about catastrophic technologies—there are compelling reasons to believe that the positivist cosmology of engineering knowledge is uniquely misleading.

2 FINITISM AND FAILURE: ON THE LOGICAL IMPLAUSIBILITY OF ULTRAHIGH RELIABILITY

Wisdom sets bounds even to knowledge.
—Nietzsche

2.1 ON THE LIMITS OF OBJECTIVITY

REASON TO DOUBT

Japan's experience with Fukushima testifies to the importance of expert reliability assessments, and the enormous trust placed in them. Of the many engineering assertions that feed into our public policy—informing decisions and shaping priorities—few are as consequential as the reliability claims made about catastrophic technologies.

As well as testifying to the importance of these claims, however, Fukushima speaks to their fallibility. Simply put, the experts who designed and evaluated the plant were wrong in their assessment of its failure behavior. And if they were wrong with respect to Fukushima, then why not elsewhere? This question is worth examining, not least because there are good reasons to imagine it has an uncomfortable answer. There are logical and pragmatic reasons for laypeople and bureaucracies to treat most engineering assertions as incontrovertible facts. This is undeniable. Yet there are equally logical reasons to believe that engineering assertions about ultrahigh reliability are exceptional in this regard and should not be treated the same way. As we will see, decades of scholarship on the integrity and provenance of expert knowledge suggest that the predicted failure rates of catastrophic technologies are much less trustworthy than they appear. In this area, if nowhere else, it is rational to distrust the experts.

To understand why this is, it helps to visit the science and technology studies (STS) literature and some of the basic epistemology on which that literature is premised.

A FINITIST COSMOLOGY

Chapter 1 spoke of a "positivist cosmology," wherein scientists and engineers are widely thought to interrogate the world objectively and definitively to uncover facts. Collins (1985) calls this intuitive and influential notion the "canonical rational-philosophical model" of technoscientific knowledge. As he himself explains, however, it holds little sway in modern epistemology.

Beginning in the mid-twentieth century, just as the first commercial reactors came online, logical philosophers such as Wittgenstein (2001 [1953]) and Feyerabend (1975), together with philosophically minded historians such as Kuhn (1996 [1962]), began to pick apart the scientific method. By demonstrating, in different ways, the logical impossibility of an "ideal experiment" or "perfect proof," they showed how even the most rigorous and objective "facts" necessarily rest on unprovable assumptions: for example, about the representativeness of laboratories or the trustworthiness of experimenters. The work of these scholars gave way to a new orthodoxy in epistemology, wherein even the most formal and rigorous knowledge claims were understood as negotiated, interpretive, and (thus) potentially value laden.

This understanding of knowledge goes by different terms, like "relativism," "extensionalism," "finitism," and "constructivism," most of which correspond with narrow distinctions in how it is conceived and invoked. For the purposes of this text, however, I will refer to it as "finitism," a term most often associated with Bloor (1976) in this context. And I will refer to the traditional understanding of science as objective and wholly rule governed, which finitism challenges, as "positivism." Diving into the nuances of finitism here would not move the argument forward very efficiently, so let us simply note that finitists do not hold that there are *no* fundamental, ontological truths about the world, as their critics sometimes claim (e.g., Gross and Levitt 1994). Neither do they suggest that one knowledge claim is as good as any other. They simply hold that all knowledge claims—even when constrained by logic, tests, and experiments—necessarily contain fundamental ambiguities, and therefore subjective judgments: an irreducibly social component of every "fact" (e.g., Bloor 1976; Collins and Pinch 1993; Kusch 2012).

By subverting claims to perfect objectivity, finitism paved the way for social scientists to study scientific and technical knowledge itself. Inspired by the philosophers, scholars from a range of social-scientific disciplines began to explore the interpretive labor involved in forging discrete and meaningful "facts" from the unruly fabric of reality (e.g., Bloor 1976; Collins 1985; Latour 1987; Stanford 2009; Sismondo 2010). In a broad movement that gradually coalesced under the banner of STS, these scholars began to map the myriad hidden choices that go into producing knowledge, demonstrating, in a series of epistemologically conscious histories and ethnographies, that the seemingly abstruse concerns of philosophers can have tangible consequences in real-world circumstances.

The wider body of STS literature has many facets, but especially pertinent to the argument that follows is a line of research exploring the production of engineering knowledge (e.g., Pinch and Bijker 1984; MacKenzie 1990, 1996b, 2001; Bijker et al. 1989; Latour 1996). In particular, a set of studies that examine the practical uncertainties that arise when engineers draw conclusions about real-world technological performance from imperfect tests and models (e.g., Wynne 1988; Pinch 1993; MacKenzie 1996a, 1996b; Collins and Pinch 1998). This body of literature makes it abundantly clear that the engineering knowledge on which we—publics, policymakers, and bureaucracies—base technical decisions has a much messier and more complex provenance than is widely supposed. Perhaps counterintuitively, however, it rarely argues that those decisions are materially poorer for the misconception.

This apparent contradiction, wherein finitist scholarship rarely takes issue with the positivist conception of knowledge invoked by decision-makers, speaks to a longstanding tension in the discipline of STS. And understanding this tension is useful to understanding why reliability assessments of catastrophic technologies deserve to be treated differently from other engineering claims. It is to this topic, then, that we now turn.

2.2 THE PROBLEM OF RELIABILITY

UNCERTAINTY AND EFFICACY

Among the early finitists, few were as radical in their skepticism of the scientific method as the Austrian-born philosopher Paul Feyerabend. Irreverent almost to a fault, Feyerabend wrote with an unreservedness that still leads many to underestimate the rigor of his thought. In his best-known work, *Against Method* (1975), he describes university science departments as sites of indoctrination and compares them to dogmatic religious orders. After his death from an inoperable brain tumor in 1994, the headline of his obituary in the *New York Times* described him as an "anti-science philosopher."

Yet even Feyerabend, bête noire of twentieth-century positivism, deferred to science when it counted. He might have excoriated medical science in his writings, but after his death, his wife often spoke about his "total confidence" in the doctors who had treated him, as well as his unhesitating deference to their recommendations (Horgan 2016; Feyerabend 1995). Her point was not that her late husband had abandoned his principles in his final days, but that the "anti-science" label was misleading. His writing supports this interpretation. When read closely, it is clear that Feyerabend, who had a background in theoretical physics, was a nuanced critic of the scientific method who believed strongly in the efficacy, if not the inviolability, of most technoscientific knowledge claims. He would likely have been no friend to the modern-day antivaxxer or climate change denier.

This is simply to say that finitism, even in its most radical forms, has never rejected the practical utility of technoscientific expertise. It is true that modern epistemologists, as well as the STS scholars who build on their insights, routinely argue that expert knowledge claims are less inviolable than they appear. (In the context of engineering, for example, MacKenzie [1996a] observes that the "insiders" who produce facts and artifacts tend to be more skeptical of them than the "outsiders" who use them, simply by virtue of being privy to the ambiguities of their production.) Like Feyerabend, however, these scholars rarely challenge the *efficacy* of those claims. All of them would argue that technoscientific knowledge is invariably more useful than its alternatives.

There is an undeniable tension here, with scholars often walking a fine line, but it is not an inherent contradiction. The intellectual understanding that medical knowledge is inherently imperfect is not incompatible with the conviction that penicillin cures better than prayer, or that physicians speak more authoritatively on health than faith healers. The discipline of STS is perennially engaged in complex internal debates about finitism's bearing on the credibility of technoscientific experts (e.g., Lynch 2017; Sismondo 2017; Collins et al. 2017; Latour 2004), but nobody in these debates endorses a posttruth society where such experts are shown no deference. All freely concede, in other words, that epistemological misgivings about proof are often inconsequential in real-world circumstances.

So it is, to return to the matter at hand, that STS studies of engineering knowledge are often comfortable arguing that publics and policymakers idealize the objectivity and inviolability of that knowledge without concluding that this idealization is dysfunctional. Given that most critiques of proof are inconsequential in real-world circumstances (and this nuance is easily lost on outsiders), it might be entirely reasonable, even from a finitist perspective, to argue that laypeople and bureaucracies should approach technological questions as if they were positivists who believed engineering assertions to be objective and inviolate.

Constant (1999) articulates the logic of this position most clearly. He accepts the argument that absolute truth is an unrealizable goal in science and engineering—all engineering facts are "immutably corrigible, hypothetical and fallible," as he puts it (355)—but he points out that technoscientific knowledge rarely has to be perfectly *true* to be *useful*. This is especially pertinent in engineering contexts, he argues, since engineers, relative to scientists, are more concerned with "application" than "discovery," and inherently less interested with what is "true" than with what "works" (352–355). Hence, he claims, the inherent limitations of their knowledge need not diminish its practical authority. "[T]here is profound difference between reliable knowledge and unreliable stuff" (335), he writes, and "most of our stuff more or less works most of the time," despite the misgivings of modern epistemologists (331). (For further exploration of this general argument, see, e.g., Petroski [2008] and Vincenti [1990].)

Constant makes a good point. After all, engineering *does* work, as our lived experience consistently affirms. Cell phones connect; bridges do not collapse; cars start. Our artifacts might not always be perfect, but longstanding experience suggests that expert claims about their properties deserve to be trusted in most practical circumstances. A society that refused to heed expert assertions that a bridge was too weak to stand, or a rocket too underpowered to fly, would quickly learn some salutary lessons.

This, in essence, explains why state bureaucracies, courts, and other rule-making bodies have long been inclined to accept engineering claims as incontrovertible facts. Real engineering practice might not perfectly match its positivist caricature, but that caricature has long contributed to a functional relationship between civil society and technological expertise.

When it comes to the reliability of catastrophic technologies, however, the conventional bureaucratic relationship to engineering knowledge breaks down. Because in this context, if in few others, there are compelling reasons to believe that the epistemological limits of proof pose real dilemmas with practical consequences.

To understand why, it helps to begin by considering the nature of reliability itself.

AN UNCONVENTIONAL VARIABLE

Considered closely, reliability is a surprisingly complicated property of artifacts. As engineers use the term today, it is usually understood to represent a discrete variable: measurable and quantitatively expressible in much the same way as an artifact's mass, density, or velocity. Yet it only really assumed this meaning in the twentieth century. (Some writers identify the V2 rocket as the first industrial system for which a reliability level was intentionally defined and experimentally verified [e.g., Villemeur 1991].) Before then, reliability was conventionally understood as a qualitative virtue: more akin to "fidelity" or "trustworthiness" than to "mass," "density" or "velocity" (connotations that still linger in nonengineering usage).[1] This unusual provenance is reflected in the fact that reliability now sits awkwardly among other engineering variables, differing from them in ways that have meaningful implications.

Of these differences, four in particular are worth noting here. The first is that, unlike most variables, reliability is a contextual property of artifacts: it must be defined in reference to agreed-upon (but always contestable and potentially changeable) measures and definitions of "acceptable" functionality (Johnson 2001, 250; Shrader-Frechette 1980, 33). Reduced to its barest form, for instance, engineering textbooks often express reliability as the frequency of failures over a given time or number of operations (e.g., Bazovsky 1961). To apply even this basic definition, however, it is necessary to specify several ambiguous and context-dependent terms. The choice of whether to define reliability in respect to time or operations, for example, is a vital but qualitative distinction. (Consider, for example, a system that performs ten million operations between failures but performs a million operations per second; or, conversely, a system that goes four decades between failures but operates only twice a decade. Each system would probably be considered reliable by one metric but unreliable by the other: the first would fail every ten seconds; the second every eight operations.) Compounding this difficulty, moreover, are ambiguities in the definition of "failure" itself. Determining when a system has failed requires that we first define what constitutes normal usage and proper operation, neither of which is straightforward in every context. In the early days of the Fukushima disaster, for example, prominent experts gamely claimed that the plant had not actually failed because it had never been designed to withstand a tsunami of such magnitude (see, e.g., Sir David King, quoted in Harvey 2011).

A second distinctive property of reliability is that, when used to express the future (as opposed to past) performance of a system—as is invariably the case in catastrophic technological contexts—then it is always, on some level, an expression of certainty or confidence (i.e., in that system's future failure rate). This point is important. With most variables, engineers can compensate for uncertainty in their measurements or calculations by hedging their numbers with tools such as error bars, which independently express their confidence in those measurements or calculations. Given that reliability calculations are already expressions of confidence, however, such hedging is nonsensical. It would be pointlessly baroque for engineers to assert that they are 99 percent certain that a rocket will launch reliably, but only 50 percent certain that this assertion is accurate.

A third notable feature of reliability is that it is a negative property of artifacts. This is to say that, unlike most engineering variables, it denotes an "absence" (i.e., of failure). ("Safety is no accident," as is written in granite outside the UK Civil Aviation Authority Safety Regulation Group headquarters [Macrae 2014, 16].) The distinction is significant because absences are famously difficult to demonstrate empirically (Popper 1959). To borrow a common formulation of this well-known problem, consider that demonstrating the presence of white swans in a given territory would simply require the observation of just one white swan, whereas demonstrating the absence of black swans would require the observation of all swans in that territory and the absolute knowledge that no swans had avoided detection. The latter proposition is inherently more difficult than the former, and it would become progressively more difficult as the territory expanded. So it is that demonstrating an absence of failure becomes ever more challenging as the required mean-time-to-failure grows larger (Macrae 2014, 16). Here, time is the "territory" in which accidents are not supposed to exist, and as it grows to billions of hours of operation, it rapidly outpaces the relative size of the observable territory (i.e., the operational hours) available to experts. Proving that a system can operate for ten minutes without failing is relatively trivial in most circumstances, but proving that it can operate for a million years between failures is not.

A final point worth noting—a corollary of the previous point—is that in most contexts, reliability is a backward-looking, actuarial property of systems. Unlike most engineering variables, it is a statistical measure of a system's relationship to the world. And, as such, it becomes empirically visible only when there is statistically significant, documented experience of that system operating in the world (i.e., service data).[2] This is simply to say that claims about reliability are usually grounded in the past, even though they routinely speak to a system's future performance. Reliability assessments of infantry rifles, for example, are essentially expressions of how often those rifles have failed in service, combined with basic *ceteris paribus* assumptions about the future (that the circumstances of rifle use and manufacture will remain constant, for instance).[3]

To summarize, therefore, let us say that "reliability"—as engineers usually understand it—is a "contextual expression of actuarially derived confidence in a putative absence." As we will see, however, these properties, in combination, have complex ramifications for finitist arguments about proof, and in catastrophic technological contexts specifically, they pose seemingly insurmountable dilemmas.

To understand why this is the case, it is necessary to consider the unique demands of catastrophic technologies.

RELIABILITY AND CATASTROPHIC TECHNOLOGIES

As we have seen, catastrophic technologies have two fundamental reliability requirements. The first is that they must be *ultrareliable*, demonstrating extraordinary mean-times-to-failure. It might be true that "most of our stuff more or less works most of the time," as Constant put it, but modifiers like "most" and "more or less" have no place in discourse about catastrophic technologies (1999, 331). The second is that their reliability must be *known*, and known *prior* to their operation. They demand what Wildavsky (1988, 77–79) calls an "anticipatory" model of safety. No organization would, or legally could, deploy a new reactor or jetliner unless experts were already satisfied that its design was ultrareliable.

Together, these two requirements—that systems have ultrahigh reliabilities, and that this be established prior to their operation—impose austere demands on expert reliability calculations, exposing them much more directly to the kinds of epistemological issues that concern finitists. The reasons for this become clear if we consider how these requirements interact with the fundamental properties of reliability as a variable, as previously described. Future chapters of this book will unpack and clarify these relationships and their ramifications in more detail, but their essence can be captured in the following four observations:

1. *The need to establish the reliability of catastrophic technologies prior to service means that experts must derive it from tests, without recourse to actuarial service data.* In this context, engineers cannot simply extrapolate failure rates from real-world performance data, as they would with most other systems; instead, they must ground their assessments in bench tests. As we will see, however, tests are always imperfect reproductions of the real world, and extrapolating from them introduces a lot of uncertainty (or a lot more than would arise from extrapolating from service data), because it forces experts to contend with questions pertaining to the relevance and representativeness of the tests themselves.

Yet . . .

2. *The need for ultrahigh reliability means that tests alone are insufficient for establishing the performance required of catastrophic technologies.* There are various reasons for this, as we will see, but the most straightforward of them pertain to time and resources. Engineers use statistical theories to determine the minimum sample size and test duration required to establish a given level of reliability in a system. At the extreme levels of reliability required of catastrophic technologies, however, these numbers become prohibitive. Even if many tests were run in parallel and were assumed to be perfectly representative of the real world, it would still take thousands of years to statistically establish ultrahigh reliability. (It would also be a self-defeating endeavor, given that any catastrophic failures these tests evinced would, in themselves, represent the kind of disaster that experts were explicitly seeking to avoid.)

Therefore . . .

3. *Since engineers cannot empirically establish ultrahigh levels of reliability with tests alone, they must combine tests of a system with theoretical models of its functioning.* To transcend the practical limits of what tests can claim to demonstrate, engineers combine the tests with elaborate theoretical representations of a system's functioning. By modeling the effects of redundancy in a system, for instance, engineers can theoretically employ test results to demonstrate much higher levels of reliability than would be possible from tests alone. In doing so, however, they make their reliability calculations dependent on even deeper levels of theoretical abstraction, for they must make complex judgments about the relevance and representativeness of their models, as well as that of their tests.

This increased exposure to questions of representativeness and relevance might be manageable in some circumstances, *except that . . .*

4. *The ultrahigh reliability required of catastrophic technologies implies a commensurately ultrahigh degree of certainty in the correctness of any judgments made in assessing them.* Recall from the previous discussion that predictive measures of reliability can be understood as expressions of confidence (i.e., that failures will not occur). It follows from this that extremely high reliability demands extremely high levels of certainty. To establish that a system is extraordinarily unlikely to fail, in other words, experts must be extraordinarily confident of the accuracy of their tests and models. (Because when looking for potential failures over billions of hours of operation, even the smallest doubt about the relevance of a model or the representativeness of a test becomes meaningful.)[4] This need for certainty has important ramifications. In contrast to Constant's portrayal of engineering as a practical discipline, it makes the task of establishing ultrahigh reliability more akin to a search for truth than a search for utility or efficacy. In this context, if nowhere else, therefore, there is little difference between "reliable knowledge" and "reliable stuff."

In light of all this, it is reasonable to assume that experts would struggle to predictively establish the failure performance of even very straightforward systems to ultrahigh levels, and catastrophic technologies are far from straightforward. Most are highly complex, consisting of many interdependent social and technical elements that interact in nonlinear ways. (As Jerome F. Lederer, the director of manned space flight safety at the National Aeronautics and Space Administration [NASA] put it in 1968: "Apollo 8 has 5,600,000 parts and 1.5 million systems, subsystems and assemblies. With 99.9 percent reliability, we could expect 5,600 defects." [Lederer 1968]) Those elements often have unusually tight design tolerances—airplane components are extremely sensitive to mass and volume, for example—which limit the scope for engineers to compensate for uncertainty with generous margins (Younossi et al. 2001). It is not unusual, moreover, for these technologies to harness uncommon materials like rare metals or innovative composites and imperfectly understood phenomena like wind turbulence or radioactivity. (Radioactivity has a poorly modeled relationship to metal fatigue, for example, especially over long periods.) And, perhaps most distinctively, many catastrophic technologies must actively negate the effects of powerful and dangerous forces, such as gravity or fission, which are inherently unforgiving of failure and require systems to do constant work simply to remain stable.

These factors exacerbate the difficulty of technological systems to the point where even relatively low levels of reliability might be seen as an achievement, let alone those required of catastrophic technologies. Witness, for instance, the extravagantly explosive history of rocketry (e.g., Swenson, Grimwood, and Alexander 1998; Longmate 1985; Clarke 2017 [1972]; Neufeld 1990) and the reliability crisis that gripped the US defense establishment during much of the Cold War (Jones-Imhotep 2000; Coppola 1984).

EPISTEMOLOGICALLY IMPLAUSIBLE

The engineering difficulty of catastrophic technologies, combined with the epistemological difficulty of predictive, ultrahigh reliability assessments, stretch the bounds of plausibility. Taken together, they strongly suggest that finitist critiques of the "perfect proof" should matter to assertions made about the reliability of catastrophic technologies, even if we agree (for the sake of argument) that those critiques are usually irrelevant to other engineering assertions. Relative to most engineering calculations, reliability assessments of catastrophic technologies are unusually dependent on simulations—tests and models, each riddled with contestable assumptions about representativeness and relevance—and, simultaneously, they are unusually dependent on the fidelity of those simulations being perfect. These relationships greatly increase the practical significance of finitist arguments about proof, especially when combined with the complexity of the systems themselves. Decades of STS scholarship suggest it should be impossible to predict the performance of a complex system over a long timeframe, to a high degree of confidence, without recourse to historical service data. The task simply requires too many subjective judgments to be made, with too much perfection, to be plausible.

Since expert understandings of *how* reliably catastrophic technologies function hinge completely on expert understandings of how those technologies function, moreover, it follows that catastrophic technologies should be unreliable for the same reason that their reliability is unknowable. Seen from a finitist perspective, their designs represent unfathomably vast networks of knowledge claims: subtle understandings of everything from the properties and behaviors of their materials in highly specific configurations and conditions, to the nature of their operating environment over long time periods, and everything in between. These knowledge claims, in turn, rest on innumerable observations, measurements, experiments, tests, and models, each ladenwith its own theories and interpretations. And any error in any step in any of these immense chains could potentially manifest as a source of failure. Insofar as it is impossible to know, with near-absolute certainty, that every knowledge claim implicit in a system is accurate, therefore, then it should be impossible to design that system to be ultrareliable.

STS scholars have traditionally striven to remain agnostic on the truth or falsity of the knowledge claims that they explore (Bloor 1976), and, like Feyerabend, they routinely defer to expert authority in practical matters. As we have seen, however, expert claims about the reliability of catastrophic technologies are functionally claims about certainty. And if there is one matter on which STS has traditionally taken a normative position, it is certainty. Indeed, the entire discipline is premised on the understanding that logic and method, even at their most rigorous, can never completely banish uncertainty.

In this context, therefore, it is reasonable, even from an STS perspective, to take a strong position on the validity of a contemporary knowledge claim. It is not only reasonable, in fact, but necessary. The assertions that experts make about the future performance of catastrophic technologies appear to be logically incompatible with an STS understanding of engineering knowledge. Everything the discipline believes about the provenance of technoscientific knowledge suggests that we should distinguish assertions made about the reliability of catastrophic technologies from other engineering assertions and doubt their plausibility. Epistemologically, it should be impossible for experts to understand a complex technological system so thoroughly, so deeply, and so perfectly as to confidently predict its failure behavior over hundreds of millions of hours of operation in a stochastic environment. By all rights, STS scholars should greet claims of ultrahigh reliability in such technologies in the same way physicists greet claims about perpetual motion.

In a world of catastrophic technologies, where the fates of major cities hinge on assertions of ultrahigh reliability, this conclusion has sociopolitical implications that are, to borrow an endearing engineering phrase, "nontrivial." From both a theoretical and a practical standpoint, therefore, it is important that we grapple with empirical evidence that appears to prove this conclusion wrong.

3 THE AVIATION PARADOX: ON THE IMPOSSIBLE RELIABILITY OF JETLINERS

The bulk of mankind is as well equipped for flying as thinking.
—Jonathan Swift

Which is now a more hopeful statement than Swift intended it to be.
—Will Durant

3.1 UNDENIABLY RELIABLE

A TRIUMPH OF MUNDANITY

In late October 2015, Norwegian Flight 7015—a Boeing 787 Dreamliner: red and white with a decal of a Scandinavian luminary on its tail—departed London en route to New York. Shortly after leaving the tarmac, its captain greeted his passengers over the public address system, wishing them a pleasant flight. He made no mention of an Atlantic storm that loomed portentously in the airplane's projected path. About an hour into the journey, as the skies began to darken, the flight ran into turbulence. A second, terser announcement instructed passengers to fasten their seatbelts. Soon afterward, the airplane's advanced in-flight entertainment system seized and had to be restarted. A small child wailed the bitter, mournful howl of a banshee heralding terrible portents.

About two movies later, the airplane touched down gently at JFK Airport—another wholly unremarkable journey, with all the romance and wonder of the modern jet age: small discomforts, trivial glitches, and officious security hurdles. It presumably flew around the storm; I was watching the movies rather than the flight tracker.

Excuse the melodrama. The point is simply to highlight the fascinating mundanity of modern air travel. The chapters that follow are punctuated by jetliner terrors and tragedies. Passages that begin with a date and a flight number invariably end with a mortality figure. When reading these stories, however, it is important to remember that such tragedies have become extraordinarily rare. The bigger picture is that every day, in almost every weather condition, private companies around the globe pack millions of people into giant machines, lift them high above the clouds, and then return them gently to Earth hundreds, or sometimes thousands, of miles from their point of departure. They have done this millions of times a year, for decades, and in recent decades they have done it with vanishingly few catastrophic failures. And—incredibly—this impresses almost nobody. The days when passengers would routinely applaud safe landings have long passed.

Understood properly, however, the wholly unremarkable, taken-for-granted safety of civil aviation is among the greatest engineering achievements of the last century—greater perhaps than the Moon landings. As outlined earlier, jetliners are inherently challenging technologies. The largest of them integrate upward of seven million components and over 170 miles of wiring. These components are highly constrained by weight and volume, often forcing engineers to work much closer to design tolerances than they would otherwise prefer. Combined, they make integrated systems of staggering complexity, which are expected to carry large volumes of flammable liquid[1] high into a stochastic operating environment that constantly threatens to punish any loss of power, control, or structural integrity. And they are expected to do this for decades, in a broad range of climates, and with minimal downtime for inspection and maintenance,[2] without ever failing catastrophically (see, e.g., Morris 2017).

As chapter 2 explained, this expectation ought to be quixotic. By all rights, it should be impossible, epistemologically, to know a system of such complexity well enough to achieve such reliability. That would require anticipating—and accurately assessing—every possible failure condition that could occur over billions of hours of enormously varied operation. The mundanity of modern air travel is fascinating, therefore, because it directly challenges this conclusion. Seemingly in defiance of some of finitism's core tenets, our experience with modern jetliners suggests that they are, in fact, ultrareliable.

KNOWABLY RELIABLE

Jetliners are not unique in being publicly understood to be ultrareliable. Indeed, it is our lived experience with most catastrophic technologies that they very rarely fail. The world's baroque atomic arsenals are yet to erupt in a single accidental missile exchange. Meltdowns—nuclear and financial—are once-in-a-generation events. And, in all cases, experts claim extraordinary reliabilities of these systems.

As we have seen, however, it is plausible to argue that the experts must be wrong with respect to most catastrophic technologies because, in most cases, our lived experiences with those technologies can be highly misleading. If a system operates for years without a major accident, that intuitively feels like compelling evidence of extreme reliability. But if that system is expected to operate for hundreds of thousands of years between failures, and there are only a modest number of instances of that system in operation, then those years of operation don't prove much. It could be far less reliable than was claimed and decades might still pass between catastrophic failures. Most catastrophic technologies are like this. They accrue service experience so slowly that the mean-times-to-failure that experts assert of them cannot be tested against empirical data. And in such circumstances, expert assertions about their reliabilities depend on abstractions—lab tests and theoretical models—the accuracy and relevance of which are open to principled critique.

As the closing chapter of this volume will explain in more detail, reactors exemplify this relationship. Even decades after the dawn of nuclear energy, the reliability attributed to even the oldest nuclear plants cannot be demonstrated statistically. In essence, this is because they are few in number and highly varied in their designs. As of 2016, there were only 444 reactors in operation worldwide (NEI 2016), no two of which were exactly alike and many of which differed very substantially from each other. In combination, these conditions are an actuary's nightmare: the small number of operating reactors limits the rate at which they collectively accrue service data, and the diversity of their designs limits the statistical relevance of each reactor's service data to that of the wider group. (Just as it would be misleading to invoke the service history of a 1972 Ford Escort to cast doubts on the reliability of a 2023 Tesla Model Y, so it would be misleading to invoke the performance of an early UK molten salt reactor to validate the reliability of a third-generation US pressurized water reactor.) The result is that, even if there had never been a single catastrophic meltdown—which, of course, there has been—statisticians would still be unable to demonstrate, actuarially, that reactors achieve the ultrahigh reliabilities ascribed to them (Raju 2016). The only way for experts to know that reliability, therefore, is through elaborate predictions, the validity of which can be challenged on epistemological grounds.

Jetliners are fundamentally different in this respect. They operate in much larger numbers than most other catastrophic technologies, and (as we will see at some length) with far less variation among their designs than is probably intuitive. Airframers often sell thousands of a given jetliner type, and airlines operate those jetliners almost continuously for years; sometimes upward of eighteen hours a day. This combination of operating volume and design commonality allows jetliners to quickly generate huge volumes of relevant service data. In 2014, for example, there were 25,332 commercial jetliners in regular service.[3] Together these jetliners averaged about 100,000 flights every single day, accruing over 45 million flight-hours over the course of the year (ATAG 2014).

Under these conditions, the service data on jetliners eventually become statistically significant, even relative to the ultrahigh levels of reliability required of them. So it is that the predicted failure performance of jetliners— unlike that of reactors and almost any other catastrophic technology—can be examined actuarily by looking at how often they have actually failed in service. Aviation experts must still assess the reliability of new designs predictively, via tests and models, as a condition of those designs being allowed to operate in the first place. But, uniquely, that reliability—and thus the accuracy of the predictive assessments—can then be examined against experience. And somewhat confoundingly, at least from a finitist perspective, jetliners in the past have performed about as well as experts predicted they would—better even.

SAFETY IN DATA

The accident statistics for modern civil aviation are truly remarkable. In 2017, for example, no passenger-carrying commercial jetliner was involved in a fatal accident anywhere in the entire world (Calder 2018).[4] None. In the course of that year, airlines moved more than 4 billion people on almost 37 million flights, with a total hull-loss rate—including nonfatal losses, cargo flights, and turboprop aircraft—of 0.11 per 100,000 flight hours (or 1 in every 8.7 million departures). To be fair, this was a record year, but broader statistics still speak almost as eloquently to the safety of modern air travel. For example, the five-year global hull-loss rate from 2012 to 2016 was only marginally less perfect, at 0.33 per 100,000 flight hours, and over the ten-year period from 2002 to 2011, there were 0.6 fatal accidents for every 1 million flights globally; or 0.4 accidents (and 12.7 fatalities) per million hours flown (CAA 2013).

The data are even more impressive when framed in relation to specific operating regimes. To date, for instance, no fatal accidents involving a UK-registered jetliner have occurred since 1989, when British Midland Flight 92 lost power and crashed outside East Midlands Airport. That makes well over three consecutive decades of catastrophe-free air travel. The US record is barely less pristine, especially relative to its much larger volume. In 2017, US-registered jetliners carried roughly 841 million people on scheduled commercial flights, amassing 17,853,752 flight hours. And, by the end of that year, it had been almost a decade since their last fatal accident (Insurance Information Institute 2018; Calder 2018; Lowy 2018; Orlady 2017).[5] Speaking in 1989, Boeing's safety manager captured the industry's already impressive reliability achievements in a colorful statistic. "If you were born on an airliner in the US in this decade and never got off," he said, "you would encounter your first fatal accident when you were 2300 years of age, and you would still have a 29% chance of being one of the survivors" (Orlady 2017, 23).[6]

Remember also that a significant percentage of the accidents in these data have little or nothing to do with failures of the jetliners themselves. They include runway incursions, terrorist attacks, pilot errors, pilot murder-suicides, and other incidents. This makes the industry's reliability record even more impressive, for, as discussed previously, reliability is a *necessary* condition for safe air travel, even if it isn't a *sufficient* condition. It is difficult to find good data on the number of aviation accidents attributable to technical failure—not least because that distinction is highly contestable (chapter 6 will touch on this in more depth)—but that number is necessarily smaller than the total number of accidents.

TRUST IN NUMBERS?
Academic observers of civil aviation, especially its critics, sometimes parse these numbers in ways that complicate the conclusion that civil aviation is safe. Fraher's (2014) broadside against US aviation regulation is exemplary in this regard. Presumably conscious that the accident data might undermine

the coming critique, she opens her book by addressing the industry's record. To this end, she equates the purveyors of aviation safety data with Wall Street analysts prior to the 2008 financial crash: calling the "apparently low aviation fatality rate" an "illusion," attributable to "luck and data manipulation" (Fraher 2014, 14–16). Her argument is misleading, especially as it pertains to the core question of this volume, but it is worth pausing here to consider its main themes. For statistics can indeed be deceptive, especially those pertaining to risk (e.g., Blastland and Spiegelhalter 2013; Power 2007), and the impressive reliability of civil aviation is integral to the argument that follows.

Fraher substantiates her skepticism on three main grounds. First, she argues that past performance is no guarantee of future safety, and might even jeopardize it. (To support this point, she cites evidence of growing complacency in the industry arising from its past success.) Second, she argues that aviation's impressive safety data tend to consider only large jetliners rather than other categories of civil aircraft, such as private aircraft, cargo aircraft, and small turboprops, which have a less robust record. And third, she argues that small numbers are deceptive because a tiny increase in the number of accidents can have a dramatic effect on the overall service record. "[A]viation quants admit the occurrence of just one accident would sway their results." She writes, noting that if the 2009 crash of Air France Flight 447 was included in the 2013 data, then fatality statistics for that year would drop "from 1 in 22.8 million to about 1 in 14 million flights: a 37 percent decline" (Fraher 2014, 16).

All these arguments contain some insight, especially in the context of Fraher's argument about regulation. For our purposes, however, none of them has much purchase.

The first is ultimately irrelevant to the central argument of this volume, which pertains to the fundamental achievability of ultrahigh reliability. It may be true, as Fraher implies, that US aviation safety is about to experience a dramatic decline due to complacency within the industry (this is actually plausible, as we will see in chapter 11). But even if aviation accidents suddenly soared for the reasons she proposes, the industry would nevertheless have proven itself capable of achieving epistemologically confounding levels of reliability over a sustained period of decades. And, as such, it would still pose a dilemma for finitists with important ramifications for technology governance. (It is probably worth noting, moreover, that critics have been making similar predictions about complacency for at least a quarter of a century without their fears ever manifesting in the statistics [e.g., Schiavo 1997; Nader and Smith 1994]).

The second argument is slightly more substantial. It is true that most civil aviation safety statistics refer only to large commercial jetliners, and the accident rate is higher in other categories of civil aviation. At the same time, however, large jetliners carry far more people than any other category of airplane, even if they constitute only about half the total departures. It also important to note that the safety record of wider civil aviation is only marginally less impressive, especially as it pertains to large jetliners in cargo roles. (FedEx's aircraft are not exactly raining from the sky.) And, crucially, it is the reliability of jetliners, specifically, that forms the crux of the discussion here. As in the previous point, if large passenger jetliners can achieve ultrahigh levels of reliability then that still poses epistemological questions, regardless of how other categories of aircraft are performing.

The third argument is just bad statistics. It is certainly true, as Fraher asserts, that small numbers can have counterintuitive implications for how we should understand the safety of jetliners or any catastrophic technology. Indeed, this is an important insight to use when reckoning with such technologies. (If the world erupted in an accidental atomic war tomorrow, for instance, then decades of zero people being killed by malfunctioning deterrence networks would become statistically meaningless in a single devastating instant.) In the context in which she is invoking it, however, the insight does not count for much.

Take, for example, her specific point that a single accident like Air France Flight 447, had it occurred in 2013, would have changed that year's fatality statistics from "1 in 22.8 million to about 1 in 14 million flights," leading to a "37 percent decline" in aviation safety. This claim isn't inaccurate per se, but it is highly misleading. This is because—somewhat ironically, given Fraher's admonition about small numbers—it overinterprets small numbers. In the parlance of risk calculation, we might say she is invoking relative risk where absolute risk would be more appropriate (Spiegelhalter 2017). This is simply to say that while "37 percent" sounds like a substantial increase, the data are always going to be "lumpy," so to speak, on a yearly basis when dealing with so few accidents every year, especially when expressed in percentage terms. If we accept that there were zero commercial jetliner fatalities in 2017, for instance, then a single fatality in 2018—expressed as a percentage change, as per Fraher's example—would represent an infinite increase in risk from the year before, while barely altering the overall level of aviation risk for the decade as a whole. While it is true that the loss of Flight 447 would have altered the statistics for 2013, therefore, it would not have meaningfully altered them for the period 2010–2020. Besides, even if we were to allow that aviation risk can be expressed in relation to an isolated year—which we should not—one fatality for every 14 million flights is still a remarkable (and epistemologically puzzling) record.

This is all to say that the data on jetliner service paint a compelling picture, even if they might be contestable at the margins. Relative to many data sets, the records on commercial flight safety are enviably comprehensive: vanishingly few departures, accidents, or fatalities go unrecorded. And while those data do not necessarily imply that civil aviation's structures and practices are beyond improvement, or that its record will not deteriorate in the future, they do speak authoritatively and affirmatively to the specific question of whether it is possible to build and assess ultrareliable jetliners. And, in doing so, they present a paradox.

3.2   A PARADOXICAL ACHIEVEMENT

PROBLEM AND PROMISE

Chapter 2 made a principled argument that ultrahigh reliability should be unachievable and unverifiable in complex technological systems. Jetliners ought to exemplify this argument. They are highly complex and inherently difficult systems. The reliability required of them is extraordinary—as demanding as that required of any system ever built, with mandated mean-times-to-failure north of hundreds of millions of hours.[7] (Chapter 4 outlines the specific requirements in more detail.) And, as with other catastrophic technologies, that performance must be established predictively, before a new type of jetliner can enter service. A finitist, STS understanding of engineering knowledge implies that these conditions should impose impossible demands on the experts responsible for achieving and measuring that performance. The task implies a depth and certainty of technical knowledge that is incompatible with the inherent ambiguities of the tests and models from which that knowledge is supposed to be derived.

So it is that civil aviation's statistically visible service record, together with the ultrahigh reliability to which that record testifies, raise academic questions with far-reaching policy implications. Contrary to core tenets of the STS literature, the history of modern jetliners strongly suggests that the ultrahigh reliabilities claimed of catastrophic technologies are not in fact unachievable or unverifiable. In this sphere at least, experts seem adept at interrogating complex systems to such an extraordinary degree of accuracy that they can accurately control, and predict, the failure performance of those systems over billions of hours of operation.

Let us call this the "aviation paradox."

This paradox is the reason why jetliners are a uniquely interesting catastrophic technology, and why the structures through which they are managed are worth exploring. It has broad implications for the governance of all catastrophic technologies, and potentially great promise. Because if aviation experts can surmount the hurdles of ultrahigh reliability in jetliners—the only catastrophic technology for which we have statistically meaningful failure data—then it seems intuitive that other experts, using analogous tools and practices, might do the same for other complex systems with similarly extreme reliability requirements. Insofar as we can successfully design and assess ultrareliable jetliners, in other words, then why not reactors, deterrence networks, drilling platforms, banking computers, or anything else?

Let us turn, then, to the question of how civil aviation experts manage the reliability of jetliners.

II CONFRONTING ULTRAHIGH RELIABILITY

Wherein it is argued:

- That the indeterminacies of assessing jetliner reliability are evident even in the ambiguities of the requirements themselves (**chapter 4**)
- That close scrutiny of reliability lab tests reveals a host of uncertainties and contested judgments, which should logically preclude assertions of ultrahigh reliability (**chapter 5**)
- That close scrutiny of reliability models similarly reveals a host of uncertainties and contested judgments, which, again, should logically preclude assertions of ultrahigh reliability (**chapter 6**)
- That the unavoidable ambiguities and indeterminacies in the processes that engineers use to understand jetliners' failure behavior mean that some catastrophic failures will always elude even the most rigorous analyses (**chapter 7**)

4 ORGANIZING AVIATION SAFETY: RELIABILITY REQUIREMENTS AND LOGICS

When we say an airline is safe to fly, it is safe to fly. There is no gray area.
—David Hinson, FAA Administrator, 1996

Recently a man asked whether the business of flying ever could be regulated by rules and statutes. I doubt it.
—Walter Hinton, 1926

4.1 RULES AND INSTITUTIONS

TYPE-CERTIFICATION

On witnessing a demonstration of the UK's first jet engine in January 1940, Winston Churchill's influential science advisor, Henry Tizard, is said to have defined a "production job" as any prototype that did not break down in his actual presence (Constant 1980, 192). In the years since then, aviation reliability requirements have become considerably more exacting. Today, the performance of civil airframes is subject to a rigorous accountability program, wherein manufacturers work closely with regulators to establish failure behaviors long into the future. The entity responsible for coordinating this work in the US is the Federal Aviation Administration (FAA). Arguably the most prominent regulator of a complex technology anywhere in the world, the FAA has many duties. Prominent among them, however, is a mandate to police the reliability of new airframe and engine designs prior to their operation.[1]

The FAA conducts this work out of three regional Aircraft Certification Offices[2] through a process that it calls "type certification." In theory, it performs this process in parallel with its European counterpart, the European Aviation Safety Authority (EASA), with both agencies issuing independent type certificates for each new airframe design. In practice, however, there is a fairly substantial division of labor, with the FAA taking the lead in certifying airframes built by US manufacturers (principally Boeing in this category), while EASA leads in certifying aircraft built by European manufacturers (principally Airbus). This cooperation is governed by bilateral airworthiness agreements and facilitated by the fact that both agencies work from standards that have been harmonized to the point of being almost identical in their wording. Together, the two agencies all but monopolize international aviation certification, with every other nation either formally or informally recognizing the type certificates of one or the other.

The extensive international deference and cooperation around type certification reflects the sheer scale of the process. Certifying a new airframe type is a forbiddingly onerous undertaking. It generates tons of paperwork,[3] takes years to complete, costs hundreds of millions of dollars, and involves skilled work by thousands of highly qualified people. For example, certifying the Boeing 777 took half a decade (1990–1995) and directly involved over 6,500 Boeing employees, as well as an unknown number of subcontractors and FAA personnel. The process utilized nine test airplanes, which collectively accumulated over 7,000 hours of flight time over the course of 4,900 test flights (NTSB 2006b, 75).

Type certification's myriad rules, protocols, and standards are codified in an extensive pyramid of guidance material, which stipulates, with ever-increasing specificity, technical requirements for the design and assessment of each element and system in a jetliner, from its structural beams to its belt buckles. At the top of this pyramid is Federal Aviation Regulation Part-25, "Airworthiness Standards: Transport Category Airplanes," commonly referred to as "FAR-25."[4] FAR-25 governs the overall design of jetliners as integrated systems. It is issued by the FAA directly, as is most of the guidance that sits directly below it (which primarily takes the form of Directives, intended for regulatory personnel and comprised of Orders and Notices, and Advisory Circulars.)[5] Many of the more specific, downstream materials to which these documents refer, however, are authored by a wide array of bodies, ranging from parallel government agencies such as the NRC or the US Department of Defense (DoD)[6] to nongovernmental organizations like the Radio Technical Commission for Aeronautics (RTCA), a volunteer organization sponsored as a Federal Advisory Committee by the FAA. Together, this great body of rules, requirements, codes, standards, best practices, and other miscellaneous guidance represents a sprawling, labyrinthine metatext of esoteric terms and layered definitions (see, e.g., appendix A of NTSB [2006b]).

PREDICTIVE AND POSITIVIST

In keeping with wider conventions around technology certification, as discussed earlier, a noteworthy feature of the guidance that governs type certification, from FAR-25 downward, is that it is predictive and positivist, largely treating each system's future reliability as an objectively quantifiable variable. This wasn't always the case. When US authorities began certifying airplane designs in 1926,[7] their requirements did not purport to calculate, quantify, or otherwise measure the reliability of the machines that they governed. Instead they established a set of proscriptive design rules for manufacturers, codified in a handbook published by the US Department of Commerce. For example, these rules specified minimum load factors for wings and mandated certain cockpit instruments. Upon designing a new aircraft for commercial use (private aircraft were excluded), manufacturers would submit blueprints to a newly created regulator,[8] which would check the design's conformance to federal mandates. If the regulator deemed the blueprints to be satisfactory, it subjected a prototype to a series of flight tests, upon the successful completion of which it issued an approved type certificate authorizing the airplane's sale and operation (Briddon et al. 1974; Komons 1978, 99).

The type-certification process retained this essential character for several decades, with its design mandates growing increasingly detailed and voluminous. Beginning in the middle of the twentieth century, however, its essential nature began to shift. The emphasis on specific design prescriptions slowly gave way to a more calculative idiom, wherein rules were framed in terms of minimum quantitatively specified reliability requirements. This shift to minimum reliability requirements represented a subtle but meaningful change in the way experts imagined, performed, and communicated the logic of type certification. It altered the definition of a "certified" aircraft from "an aircraft shown to have satisfied every design requirement" to being "an aircraft shown to have a specified failure performance." In doing so, it premised the whole process on the ability of experts to interrogate designs and accurately quantify their failure behavior.

There were many reasons for this transition, including the aforementioned reconceptualization of reliability (from virtue to variable) in engineering more broadly, together with wider structural incentives toward quantification (which will be outlined in chapter 12). Internally, however, the US aviation community usually explains the shift in one of two ways, which are distinct but not mutually exclusive. The first explanation construes it as a response to the growing complexity of airplanes. By this view, the FAA started calculating the reliability of systems because their designs were becoming too elaborate to judge qualitatively. "As the number, criticality, complexity, integration, and number of parts of aircraft systems increased," explains an FAA publication, so "the combinations of conditions and events that a design must safely accommodate became more difficult to effectively judge by qualitative means alone" (FAA 2002c, 23). The second explanation construes the shift as an attempt to better accommodate innovation in the industry. By this view, the need for manufacturers to meet proscriptive design requirements was constraining their ability to reimagine airplane systems and architectures. Framing regulations around minimum reliability requirements was seen as a way of mitigating this. In theory, at least, it allowed engineers to build airplanes however they chose, so long as they could demonstrate that their designs met a satisfactory level of failure performance.

The exact mechanisms by which manufacturers are supposed to demonstrate conformance with type certification's reliability requirements vary. (The process treats "systems" differently from "structures," for instance.)[9] In most instances, however, the procedures are flexible, such that regulators will consider evidence from a vertiginous range of analytical tools.[10] The competing logics of these tools are highly esoteric, but they all draw on and manipulate data derived from the same foundational processes.

Ostensibly at least, these processes are the same for jetliners as they are for reactors and most other catastrophic technologies. This is to say that when stripped to their barest fundamentals, they involve two basic steps. The first is *testing*, wherein experts use controlled environments to empirically examine the failure performance of a jetliner's individual elements. As we will see, however, even the most extensive and idealized testing regimen could never demonstrate the levels of reliability that jetliners require;

it would take a prohibitively long time (potentially thousands of years) and the cost would be astronomical. What cannot be demonstrated empirically, therefore, must be demonstrated in principle. Hence the second step: *modeling*, wherein experts integrate their test results into representations of the wider system architecture and invoke those representations to demonstrate much higher reliabilities than could be established with tests alone.

Subsequent chapters of this book will examine these processes and their applications in more detail. Before then, however, it is worth pausing to consider the nature of the reliability requirements themselves. Just as tools for interpreting reliability data can only be as good as the data they interpret, so procedures that use reliability metrics to govern airframes necessarily hinge on the validity of their requirements. It doesn't much matter if experts are failing to measure a system's performance accurately—in other words, if the level of performance they are trying to ensure is insufficient to achieve their required ends.

4.2 RELIABILITY TARGETS

TWO NUMBERS

Perhaps the most interesting aspect of the FAA's turn to quantitative reliability requirements is how imprecisely it articulates the levels of reliability that are needed. Broadly, the failure performance that the FAA requires of new jetliners is usually expressed in terms of one of two numbers: one in ten million (10^{-7}), and one in a billion (10^{-9}) (FAA 1982, 2002b, 2002c). These numbers—10^{-7} and 10^{-9}—frame the entire type-certification process and are essential to the positivist vision of safety promulgated by the FAA and other aviation bodies around the world (e.g., Lloyd and Tye 1982). Examined closely, however, their underlying rationales are ambiguous in ways, and to degrees, that shed light on the viability of legislating for extreme reliability.

Let us consider each number in turn.

ONE IN TEN MILLION (10^{-7}) The first number, 10^{-7}, arguably represents the most fundamental goal of type certification, which is to ensure that the probability of a serious accident is no greater than "one in every ten million hours" of operation (FAA 2002b, 5). This figure, then, is the reliability that is officially required of a jetliner as an integrated system, and its origins and logic are ostensibly straightforward. Formally established in 1982, it was simply an expression of the frequency of accidents per departure at the time. As such, it represented an implicit understanding that this accident frequency needed to remain constant, if not decline, if civil aviation were to remain credible among publics and policymakers (FAA 2002c, 23). On close inspection, however, it is probably fair to say that the calculation by which the FAA arrived at this number was not the most rigorous or extensive ever performed by a federal regulatory agency.

Its rationale hinges on at least two highly questionable premises. The first is that only one in every ten accidents is caused by a reliability issue. The accident rate at the time when the number was established was actually calculated to be about one in every million hours (thus 10^{-6}), but since only one in every ten of those accidents was attributed to system failure, the probability of a reliability issue felling an airplane was calculated to be $10/10^{-6}$ or 10^{-7} (FAA 2002c, 23). This "one in ten" attribution is highly problematic, however. Of the remaining nine accidents, most were attributed to human error, and, as accident theorists have long maintained, the distinction between "human error" and "technological failure" is blurry at best (e.g., Reason 1990; Perrow 1983). ("There is no problem so complex that it cannot simply be blamed on the pilot," as an old industry adage puts it.)

The second questionable premise in the FAA's adoption of the 10^{-7} number is the assumption that a stable rate of accidents per departure would remain acceptable as the absolute number of flights (and thus the absolute number of accidents) increased dramatically. This is problematic because, as we have seen, public opinion on aviation safety appears to be more sensitive to the *absolute* frequency of accidents than it is to the *relative* frequency. The number of accidents per year tends to loom larger than the number of accidents per departure, in other words. And since the number of departures per year has risen steadily since 1984, it is highly doubtful whether maintaining the same number of accidents per departure (i.e., the 10^{-7} reliability target) would be deemed acceptable today, given that it would represent a very considerable spike from current accident levels. (In fact, the 10^{-7} number was already outdated in this regard even at the time it was adopted, having been lifted from an earlier British Civil Airworthiness Requirement [FAA 2002c]). Fortunately, the number of accidents per departure has dropped steadily since 1982, such that modern jetliners now far outperform the 10^{-7} requirement (Flight Safety Foundation 2018).

ONE IN A BILLION (10^{-9}): THE "NINE 9s" The 10^{-7} reliability target is problematic, therefore, but it is arguably less important to the certification process than the second number: 10^{-9}. This is because type certification's standards are not really framed around the jetliner as an integrated system so much as they are framed around its individual systems and assemblies. (As the NTSB [2006b, 90] puts it: "[A]irplane-level risk and hazard analyses are neither required by certification regulation nor recommended by FAA advisory materials.") Rather than assessing the reliability of a whole jetliner, in other words, regulators assess the reliability of its essential elements: the operational unit of the certification process being the "flight-critical system," (which I will henceforth refer to as "critical systems") (FAA 2003).[11] Critical systems are the subassemblies of a jetliner that would jeopardize the plane if they failed—the flight controls or landing gear, for instance[12]—and it is to them that the 10^{-9} figure refers. It is widely held that airframers must demonstrate for each critical system a mean-time-to-failure higher than a billion hours of operation.

This number, 10^{-9}—often informally referred to as the "nine 9s" since it can be expressed as a reliability of 0.999999999, where 1.0 would represent perfection—appears frequently in the discourse around type certification. By most accounts, it is the linchpin of the process, representing the key reliability metric that experts are striving to achieve in their designs and validate in their assessments. It should be somewhat surprising, therefore, that the FAA never explicitly defines the 10^{-9} reliability target. The literature around certification offers competing rationales for it, often in the same documents and sometimes even on the same page.

At the most fundamental level, these rationales derive the number from one, or often both, of two distinct certification requirements, each of which, again, rests on problematic assumptions. The first derives it directly from the requirement outlined previously: that airplanes should crash no more than once in every 10 million (10^{-7}) hours. This rationale assumes that every jetliner has exactly 100 critical systems, each of which therefore has to be 100 times more reliable than the reliability required of the airplane itself.[13] The essential math here is relatively simple. If there are 100 sources of potential failure that could cause an airplane to crash in any given hour, then the probability of any one of those faults occurring has to be 100 times lower than the reliability required of the airplane itself ($100/10^{-7} = 10^{-9}$). But deriving 10^{-9} from the 10^{-7} requirement in this way introduces hidden complications. As outlined earlier, for instance, the 10^{-7} requirement itself rests on ambiguous foundations; and insofar as it inadequately expresses the reliability required of each jetliner, then 10^{-9} will inadequately express the performance required of its critical systems. It is also far from clear that modern airplanes in fact have exactly 100 critical systems, not least because the distinction between "critical" and "noncritical" systems is highly subjective. The FAA itself has called the distinction "necessarily qualitative" (FAA 1988, 7), and found that critical systems are "not consistently identified" (FAA 2002a). Others have argued that the distinction is almost meaningless, since almost any component or system can be critical in the right (or wrong) circumstances (e.g., Perrow 1999; Leveson et al. 2009; Macrae 2014, 8). (When a jetliner leaving Boston crashed in 1960, for example, investigators found that a faulty seat-adjustment pin had caused the pilot to make an "unintended input" [Newhouse 1982, 94–96]).

The second, more common rationale given for the nine 9s is derived from a requirement in FAR-25 that catastrophic failures be "extremely improbable," defined, elsewhere in the guidance, as meaning "not anticipated to occur during the entire operational life of all airplanes of one type [its fleet life]" (FAA 2002b, 9). By this rationale, the 10^{-9} reliability target is understood as the level of performance needed to satisfy this requirement. Again, however, deriving the number from the rule requires some questionable assumptions, not least because any calculation of the reliability required to make catastrophic failures "extremely improbable" by the definition given here necessarily hinges on both (1) how many planes of each type will enter service [i.e., the fleet], and (2) how long that service will last [i.e., the life]. Neither of these variables is defined in the guidance, and both have changed meaningfully over the last forty years.

Given this definitional ambiguity, it is probably unsurprising that when secondary literatures make more detailed attempts to justify the nine 9s, they tend to vary widely in their reasoning. Such efforts usually combine the two rationales: constructing an explanation that defines "fleet life" in ways that make the definition of "extremely improbable" match the 10^{-7} requirement, and then positing 100 critical systems (or failure modes) to arrive at 10^{-9} for each. Lloyd and Tye (1982), a respected study about aviation regulation, exemplifies this pattern. To arrive at the nine 9s, its authors posit a fleet of 200 aircraft, each flying 50,000 (5×10^4) hours before retirement, thereby creating a total fleet life of 10,000,000 (10^7) hours. They then suppose 100 sources of catastrophic failure in each aircraft and implicitly equate that to

100 critical systems, each requiring a mean-time-to-failure of 10^9 hours (Lloyd and Tye 1982, 37). As if to underline the interpretive flexibilities of this calculation, however, the authors then—on the same page(!)—offer an incommensurable justification of the 10^{-7} requirement: positing a fleet size of 100 aircraft instead of 200 and a life of 90,000 hours instead of 50,000.[14]

Another formulation of the same calculation was offered to the author by an FAA chief scientific and technical advisor in 2005:

> The one-in-a-billion figure comes from estimating the fleet life for one aircraft model. . . . One operational year is close to 10^4 operational hours (it is around 8500 hours), assuming 24-hour operations (many aircraft operate two-thirds of the day, every day, almost year-round), 30-year service life gives you 3×10^5 for an aircraft-lifetime, and 3,000 aircraft in the fleet . . . you get 9×10^8, which is close enough to 10^9 operational hours per fleet-lifetime.[15]

Consider the variations between these various accounts. The 1982 formulation imagines a service life for each airframe of 50,000 hours (or 90,000) and a fleet size of 200 airplanes (or 100), while the 2005 formulation posits a service life of 300,000 ("3×10^5") hours and a fleet size of 3,000 airplanes. These are very meaningful differences. Where the first concludes that a specific airplane type will accrue 10 million flight-hours during its lifetime, the second concludes that it will accrue a billion. In doing so, moreover, the latter reaches 10^8 without accounting for the fact that airframes have multiple critical systems. Upon being alerted to this apparent oversight, the same correspondent reflected candidly on the changing nature of the industry. "The [nine 9s] figure was derived at a time at which no model [of airframe] was expected to be in service 30 years, or [to be manufactured] in multiples of a thousand. . . . Maybe we should go to 10^{10}."[16]

Maybe they should.

FIGURES OF MERIT?

In light of the variance and ambiguity evident in the definitions given here, it is difficult to ascribe much rigor or agency to the quantitative targets at the heart of type certification. The seemingly crucial standards against which regulators ostensibly measure the reliability of airframes are self-evidently social constructions: rhetorically compelling, perhaps, but not mathematically meaningful. We might think of the nine 9s as a "Goldilocks number": suitably impressive without being inconceivable. It is worth noting that the same "one in a billion" figure appears in other high-profile contexts where the public demands extreme levels of certainty. It is the oft-stated reliability of DNA matching, for instance, and is similarly problematic in this context (Lynch and Cole 2002).

The less-than-rigorous foundations of certification's reliability requirements are not lost on experts themselves. One uncommonly forthright correspondent—an experienced engineer once employed by a leading manufacturer—described the nine 9s as "fatuous nonsense . . . designed to gull the public into believing that someone is actually producing a figure of merit."[17] The FAA presumably disagrees on the specific question of fatuousness, but it too sometimes will concede the broader point in its more private and esoteric discourse. In supplementary literature accompanying a redraft of the guidance that explains the numbers (FAA 2002b), for instance, it responds to what it describes as "misinterpretation, confusion, and controversy" regarding their application. In an almost parenthetical passage, tonally incongruous with the analysis that precedes it, the regulator clarifies that the numbers are intended to be "guidelines" only and stresses that they cannot replace or supersede engineering judgment when making airworthiness determinations (FAA 2002b, 25).

These kinds of caveats are especially common in the literature around certification's implementation (as opposed to its purpose). Here, it is relatively easy to find language that suggests a nuanced, practical understanding of technology assessment that is more congruent with finitist accounts of engineering than with its positivist public image. The FAA refers to key variables as "somewhat arbitrary" (e.g., FAA 2002b, 5; 2002c, 24), for example, and routinely flags the limitations of quantitative analyses (e.g., FAA 1982, 2; 2002b, 7, 25; 2002c, 23, 25). A 1983 investigation by the National Academy of Sciences (NAS) really grasps the nettle, bluntly stating that "the determination of design and engineering adequacy and product safety [in aviation] cannot be legislated in minute detail" (NAS 1980, 23). In such literature, numerical values are often "assigned" rather than "calculated," and assessments are more often required to be "convincing" rather than "correct."[18]

Such comments might look innocuous, but they have far-reaching implications for how we—publics, policymakers, and laypeople of all kinds—should construe aviation safety in general and the type-certification process in particular. The judgments they acknowledge speak to a tension between certification's messy realities on one hand, and its portrayal as a set of exacting requirements that airplane manufacturers must satisfy on the other. They also point to a dilemma. Because if experts cannot precisely define the reliability required of a system, then determining whether the reliability of that system is satisfactory cannot be a wholly objective process. And if assessing the extreme reliability of a critical system cannot be a wholly objective process, then, as we have seen, it places seemingly impossible demands on the experts' subjective decision-making.

4.3   PRACTICAL DILEMMAS

THE CHALLENGE OF AMBIGUITY

Insofar as we construe type certification as a set of exacting requirements that airplane manufacturers must satisfy, then it is tempting to imagine that its flexibilities must diminish the difficulties of satisfying those requirements (as presumably, it is easier for manufacturers to claim compliance with ambiguous standards that bend accommodatingly to interpretation). And in many catastrophic technological contexts, this might be true. As we have seen, however, civil aviation experts are in a unique position stemming from the fact that their reliability claims will eventually be tested against real, statistically significant service. In this domain, if in few others, they cannot navigate their regulatory obligations by exploiting ambiguities in the rules, because their exploits would become obvious when planes failed.

This means that airframers really do have to design and build ultrareliable jetliners, irrespective of the certification bureaucracy. The nine 9s might be a social construction but it is far from being an overestimate. Jetliners undoubtedly *do* have critical systems on which their safety depends, and meeting modern expectations about aviation safety undoubtedly *does* hinge on those systems functioning for billions of hours between catastrophic failures. (Indeed, as we saw earlier in this discussion, parsing the FAA's numbers suggests that the failure performance required of such systems is, in practice, considerably higher than what certification formally requires.) In these circumstances, it is reasonable to imagine that manufacturers would welcome well-defined requirements, and any ambiguities in those requirements are better understood as a source of difficulty rather than opportunity.

Seen in this light, the vagaries of type certification's reliability targets exemplify both the finitist case against ultrahigh reliability assessment and the finitist paradox of aviation safety that challenges that case. The reliability of a jetliner hinges on experts' ability to know the behavior of its most critical systems to an extraordinary degree of accuracy and confidence. Yet those experts must establish their knowledge with reference to fundamentally ambiguous metrics, based on dubious foundations that cannot be applied without subjective judgment and interpretation. Jetliners are demonstrably reliable, however (or have proved to be so in the past), so experts must be making those judgments and interpretations with a degree of accuracy that is incommensurate with the rigor of their rules and metrics. The question, therefore, is *how* they do this.

To begin to answer this question, we must look at the practical work of certification in action. Let us now turn to examining one of its most fundamental practices: testing.

5 WHEN THE CHICK HITS THE FAN: TESTING AND THE PROBLEM OF RELEVANCE

Not everything that counts can be counted.

—William Bruce Cameron

Experience acquired with turbine engines has revealed that foreign object ingestion has, at times, resulted in safety hazards. Such hazards may be extreme and
possibly catastrophic involving explosions, uncontrollable fires, engine disinte-
gration, and lack of containment of broken blading. . . . While the magnitude of
the overall hazards from foreign object ingestion are often dependent upon more
than one factor, engine design appears to be the most important.

—FAA Advisory Circular (FAA 1970)

5.1 BIRD-STRIKE TESTING

A PERSISTENT HAZARD

On October 4, 1960, Eastern Airlines Flight 375 struck a flock of starlings
six seconds after taking off from Boston. The birds destroyed one of the
Lockheed Electra's four engines and stalled two others. Starved of power at
a vital moment, the injured aircraft yawed steeply to the left at 200 feet and
plunged headlong into the shallow green water of Winthrop Bay. Sixty-two
people lost their lives.

Flight 375 was far from the first airplane to be brought down by wildlife.
Orville Wright, ever the pioneer, claimed the first documented "bird strike"
in 1905, just two years after he made the first powered flight. (It could not
have come as a huge surprise; his diaries record that he was chasing a flock

FIGURE 5.1
US Airways Flight 1549 landed in the Hudson River, January 15, 2009. *Source:* NTSB
(2009, 5).

of birds around a cornfield at the time.) The first reported bird-strike fatality
came seven years later, in 1912, when Cal Rodgers—a student of the Wrights
and the first man to fly across the continental US—caught a seagull in his
flight controls and plummeted into the Californian surf (Thorpe 2003).

Regrettably, moreover, Flight 375 was far from the last airplane to be
brought down by wildlife. The years since the tragedy have failed to resolve
what continues to be a problem. In January 2009, for instance, US Airways
Flight 1549, an Airbus A-320, only narrowly avoided a similar catastrophe
when a bird strike forced its pilot to ditch in the Hudson River next to mid-
town Manhattan (figure 5.1). Nobody died on that occasion, but aviation
insiders were quietly astonished by the escape. Water landings are rarely so
successful (Langewiesche 2009b). Globally, bird and other wildlife strikes
killed more than 255 people and destroyed over 243 aircraft between 1988
and 2013 (FAA 2014, 1).

Birds are not the only wildlife threat to jetliners. Any creature that flys in
the air or wanders onto runways is liable to be struck directly or sucked into
an engine. The FAA Office of Airport Safety and Standards has published a

table of animals struck by civil aircraft in the US between 1990 and 2013 (FAA 2014, 37–56). A veritable menagerie of American fauna, it lists forty-two species of terrestrial mammal and fifteen species of reptile, including 978 white-tailed deer, 443 coyotes, 243 striped skunks, 229 black-tailed jackrabbits, 19 alligators, 13 painted turtles, 5 moose, 3 horses, and 1 black bear. Elsewhere, there are even reports of at least one "fish strike"; albeit coincident with an osprey (Langewiesche 2009a). (Not astoundingly, there appears to be no established audit capacity for fish strikes.)

Such incidents can sometimes threaten the airplane, especially if they occur while it is taking off, but birds are by far the most prominent hazard. Bird strikes constitute over 97 percent of all reported wildlife strikes, with over 11,000 reported in 2013 alone (FAA 2014, x). The costs of these incidents are substantial. Birds destroyed over twenty US commercial aircraft between 1960 and 2003 (Bokulich 2003). In a 2014 report, the FAA estimated that they cost American carriers somewhere between $187 million and $937 million in damages annually, as well as between 117,740 and 588,699 hours of unintended downtime (FAA 2014, xi).[1]

Airports employ an innovative array of tactics to keep birds away from jetliners, but these are rarely more than marginally successful. Evolutionarily well designed for negotiating high fences but not for avoiding fast objects,[2] birds are often attracted to the open spaces around runways, where they then struggle to avoid the air traffic. Groundkeepers try to scare them off with a spectrum of creative measures—from plastic hawks and rubber snakes to lasers and pyrotechnics—but birds are not easily cowed, and many become inured to even the most vigorous attempts at intimidation. Scarecrows get claimed as nesting places; noise generators often double as popular perches.

Rather than bet too heavily on controlling the birds themselves, therefore, the onus is on the airplanes, which must be bird-resilient by design. This requirement falls most directly on the engines. Birds can damage any leading part of a jetliner, but the engines are most vulnerable for several reasons. They are more likely to be struck because they inhale so aggressively. The strike area that they present consists of precision machinery revolving at high speeds and pressures. (Protective grills over engine mouths are an engineering dead end. Any grill that was strong and dense enough to withstand birds at high speeds would occlude airflow to the turbines and pose its own risk of being smashed into the blades.) And their operation is safety crucial, especially during takeoff, when power demands are higher and pilots have less altitude with which to work.

For this reason, the FAA's engine certification requirements outline a range of mandatory bird-strike tests. These tests represent only a small fraction of the testing regimen involved in certifying an engine or airframe, but exploring them in depth helps illustrate a dilemma common to all tests, and in doing so illuminates the nature of reliability engineering itself.

BOOSTING ROOSTERS

In principle at least, the FAA's bird-strike tests for new engine designs are as straightforward as they are dramatic.

Engineers prepare by firmly mounting an engine onto an outdoor stand, where it is already an impressive spectacle. The largest engines—which weigh over eighteen tons and cost more than their weight in silver[3]—have mouths that yawn over thirteen feet (four meters) across. Inside are huge turbines with teeth of graphite and titanium, all balanced so delicately that a slight breeze will spin them on the tarmac like a child's paper windmill. Designed to operate at over 2,500° Fahrenheit [1,371° Celsius], well above the temperature at which most alloys melt, each blade in these turbines represents the very forefront of materials science.

With everything in place, the test begins. Engineers gradually apply power to the engine, and its turbines begin to turn. The giant fan blades spin faster and faster until the engine reaches maximum climbing speed, where it roars like a furious kraken and blows like an uncorked hurricane. At this point, the most powerful engines are gulping over 1,900 gallons of fuel an hour to produce up to 127,900 lbf (569 kN) of thrust, and the tips of their largest turbines are moving at close to the speed of sound. And then, into the maw of this howling monster—this technological jewel, the billion-dollar product of years of work by thousands of experts—they launch an unplucked, four-pound chicken.

Strictly speaking, the bird does not have to be a chicken; any bird of equivalent mass will do. And for the engine to meet its certification requirements, it must perform similar feats with other birds representing different weight categories. Along with the single large bird, represented by the chicken (4 pounds [1.81 kilograms], or 8 pounds [3.63 kilograms] for very large engines),[4] the regulations require a volley of eight medium-size birds (1.5 pounds [0.68

kilograms]), each fired in quick succession, and a further volley of sixteen small birds (8 ounces [85 grams]). (Many more will be launched in precertification tests [Bokulich 2003]). All are fired from a compressed-air cannon at takeoff speed, which is usually around 200 knots (230 miles per hour) (FAA 2000). (The US Air Force, for its part, tests its much-faster-moving airplanes with a sixty-foot cannon capable of firing a four-pound feathered bird, headfirst, at over 1,000 miles per hour. They are said to call it the "Rooster Booster.")

The pass/fail criteria are broadly the same in each test. If the turbine "bursts" (i.e., releases blade fragments through the engine cowl) when the chick hits the fan, or if it catches fire, or cannot be shut down properly, then the engine fails the test. The tests with smaller birds are more demanding. In these, engines also fail if their output drops by more than 25 percent or if they fail within five minutes of being struck (FAA 2000). (Although a drop below 75 percent is acceptable, so long as it doesn't exceed three seconds.)[5]

Once experts deem an engine to have passed the tests, the FAA holds that it has proved its ability to safely ingest all the birds that it will encounter in service. So far as the regulator is concerned, it has demonstrated its resilience to bird strikes, which cease to be a meaningful reliability concern.

Behind this technocratic surety, however, lie complex debates about what the tests actually reveal, as well as meaningful uncertainty about the confidence to be gleaned from passing them. Take, for instance, the pass/fail criteria outlined here. Although they may look relatively concise on the page, these criteria are less straightforward than they appear, as they are framed by multiple caveats, the application of which demand complex rulings. If two birds in a volley strike the same fan-blade, for instance, then the test can be deemed to be excessively challenging. Conversely, if the birds do not strike a predefined point of "maximum vulnerability" on the fan blades, then the test can be deemed insufficiently challenging. Such debates are worth examining, not because (or not *only* because) bird-strike tests are important in themselves, but because they offer generalizable insights into the wider epistemology of testing and its role in assessing the reliability of catastrophic technologies.

To understand the issues involved and their significance, it helps to begin by revisiting the finitist arguments about the limits of proof (see chapter 2) and elaborating on their application to technological testing.

5.2   REPRESENTATIVENESS

THE PROBLEM OF RELEVANCE

Bird-strike tests are what engineers sometimes call "proof tests": their purpose, as Sims (1999, 492) puts it, is "to test a complete technological system
under conditions as close as possible to actual field conditions, to make a projection about whether it will work as it is supposed to." These projections are
important; experts bet lives on their accuracy. But they are also problematic.

As we have seen, the relationship between tests and the phenomena they
reproduce has long posed dilemmas for those who seek to establish the validity of engineering knowledge claims. This is because tests always involve
some degree of extrapolation. Engineers must infer from them how the
technology will perform in practice, and such inferences inevitably hinge on
judgments about the tests' representativeness. This is to say that the validity
of a test hinges on how accurately it reproduces the real-world conditions it
ostensibly simulates (Pinch 1993; MacKenzie 1996b).

Experts cannot infer much about how a technology will behave in operation, in other words, unless they are confident that their tests *adequately
represent* the actual conditions under which that technology will operate. To
achieve this, they must either (1) design tests that simulate the real world
accurately enough that the phenomena being investigated (such as an
engine's bird-strike resilience) will present identically in both the test and
the world; or (2) understand the meaningful differences between their test
and the real world accurately enough to interpret their results in a useful
way. Since even the most realistic tests always differ from the real world
to some degree, however, satisfying either condition requires that experts
determine which differences are meaningful and which trivial, but such
determinations are inherently open-ended (see, e.g., MacKenzie 1989).

Reduced to its essence, the dilemma here is that there are a potentially
infinite number of ways that a lab test might not match the real world, in
all its nuanced and complex messiness, but testers can only hope to control—or even recognize—a finite number of variables (hence the term "finitism"). Scholars often refer to this dilemma as "the problem of relevance"
(e.g., Collins 1982, 1988), but it goes by a variety of other names. Latour and
Woolgar (1979) call it the problem of "correspondence"; and when Vincenti
(1979) speaks of "the laws of similitude," Pinch (1993) of "projection," or
Kuhn (1996 [1962]) of "similarity relationships," they are invoking the same

underlying problem. Every FAA certification test necessarily grapples with this conundrum, but the debates around bird-strike testing amply serve to illustrate its nature and implications.

"CONSISTENT WITH SERVICE"

To understand the relevance debates around bird-strike tests, it helps to begin with the data on which the tests are premised. The FAA's current tests grew out of recommendations made in a 1976 National Transportation Safety Board (NTSB) report. The report, which examined a bird strike incident involving a DC-10 out of JFK Airport, suggested that all future engines be subject to "strike tests" using birds of sizes and numbers that were "consistent with the birds ingested during service experience of the engines" (FAA 2000). To satisfy this recommendation, the FAA set out to determine the nature of real-world bird encounters with a detailed study of real bird-ingestion incidents. It collected information on the types, sizes, and quantities of the birds involved, along with their effects on different engines.

These historical data became the basis for its bird-strike certification requirements. They were used to establish the conditions under which engines would be tested, the properties that tests would seek to examine, and the performance criteria that testers would be required to demonstrate. Over time, however, many expert observers have questioned their accuracy and appropriateness. There are too many such questions to fully document here, but a few prominent issues serve to illustrate the wider debate.

One significant cause for doubt arises from ambiguities in the collection process itself. At the most basic level, collecting bird-strike data involves counting and identifying birds killed by airplanes. This work is possible because the majority of bird strikes occur when aircraft are close to the ground (either taking off or landing),[6] meaning that investigators can gather remains from the runway. Not all strikes occur at ground level, however, and even a significant number of ground-level strikes go unrecognized or unrecorded (reporting is voluntary, and birds often pass through engines without anyone noticing), so the data are necessarily incomplete (Cleary, Dolbeer, and Wright 2006). (Over the period from 2004 to 2008, for instance, the FAA estimates that only about 39 percent of commercial aviation wildlife strikes were recorded, but that figure is up from about 20 percent in the 1990s [FAA 2014, 56]). This incompleteness is compounded by the fact that even recognized strikes present data-collection challenges. As might be imagined, the experience of being sucked through a roaring turbofan engine is traumatizing for birds. It turns them into a pulpy mess, known in the business as "snarge," which spreads over a wide area. This makes individual birds difficult to find, count, and identify by species (and hence by vital characteristics such as mass) (MacKinnon, Sowden, and Dudley 2001).[7]

Further doubts arise from *ceteris paribus* concerns. This is just to say that some experts doubt whether historical bird-strike data, even if they were complete and accurate, would adequately represent future (or even present) conditions. Critics have questioned the FAA's extrapolation from old to new engines, for instance, suggesting that data accumulated from incidents with early engines (and engine configurations) may not be applicable to newer ones (e.g., FAA 2014, 2). Others have pointed to changing bird populations. Ingestion standards are largely based on the assumption that bird numbers and distributions will remain constant over time, and yet the FAA itself has found that both are changing in ways that increase aviation hazards (FAA 1998; 2014, x). Bolstered by wildlife agencies, climate change, and shifts in farming techniques, many of the species that most threaten airplanes—such as gulls and pigeons, which feed on waste, and turkey vultures, which eat roadkill—have flourished in recent decades (Barcott 2009). Of the fourteen species of large-bodied birds in North America, for instance, thirteen experienced significant population increases over the thirty-five years from 1968 to 2003 (Dolbeer and Eschenfelder 2003). Perhaps most notably, the US-Canada goose population rose from 200,000 in 1980 to 3.8 million in 2013 (FWS 2013; Eschenfelder 2001). Gull and cormorant populations around the Great Lakes have similarly expanded, with the latter rising from just 6 nesting pairs in 1972 to over 230,000 by 2020 (Drier 2020). In line with these shifting demographics, the number of strikes reported by US airports increased significantly over the same period (FAA 2014; Barcott 2009).

Even if we assume that the data informing bird-strike tests are accurate, complete, and consistent, however, questions remain about the ability of testers to represent those data adequately. So a second set of debates surround the representativeness of the tests themselves.

RECREATING STRIKES

In many respects, the FAA is extremely cautious about the authenticity of its bird-strike tests. For example, it prefers using freshly killed birds to birds that were previously frozen because the former are considered more realistic. (Thawed birds can be dehydrated, and incompletely thawed birds can contain dense ice particles [FAA 1970, 8].)[8] This concern for authenticity is less fastidious in other respects, however, and critics have raised doubts about a wide range of test parameters. These concerns are too numerous to be outlined here in full, but highlighting a few paints a useful picture of the complications involved. To this end, therefore, let us briefly visit the debates around four distinct test parameters: size/mass, species, quantity, and speed.

SIZE/MASS   As outlined earlier in this chapter, the certification standards require three sets of bird-strike tests, each with a different category of bird. These categories are labeled in reference to their size—large, medium, and small—but are defined by their mass. The "large" birds are either 4 pounds (1.8 kilograms) or 8 pounds (3.6 kilograms), depending on the size of the engine, the "medium" birds are 1.5 pounds (0.7 kilograms), and the "small" birds are 3 ounces (85 grams). The different categories are intended to represent the various species of bird that an engine might inadvertently swallow, but some experts question their correspondence with the birds that actually make it into engines.

Some birds that engines regularly encounter fit neatly into the FAA's categories. The small (3-ounce) bird is about the mass of the European starling, for example, which frequently falls prey to civil aircraft. And, although chickens themselves are rarely sucked into engines outside of laboratories, the large (4- or 8-pound) bird that they represent is about the same mass as various waterfowl, such as gulls, which are often found near runways. The medium (1.5 pounds) category, on the other hand, is less representative. Barred owls and red-shouldered hawks are approximately this mass, but aircraft rarely ingest either. Ducks, meanwhile, are a common engine ingestee, but they tend to range from 2 to 3 pounds (0.9 to 1.4 kilograms), thus falling between the test categories (Eschenfelder 2000).

Such discrepancies are more significant and contentious than might seem intuitive. Take, for example, a dispute that arose during the FAA's attempts to harmonize its standards with its European counterpart, the Joint Airworthiness Authority (JAA) (now the European Aviation Safety Authority [EASA]) (FAA 2000). The JAA had concerns that stemmed from tests that it had reviewed of an engine with fan blades made from a novel material. The tests had revealed something surprising. The new blades behaved similarly to the old ones when faced with the volley of "medium" birds and the single "large" bird. When faced with a bird of a size between "large" and "medium," however, they were found to be only "marginally equivalent" to previous designs, displaying an "inferior level of robustness" (FAA 2004). This led the European regulator to conclude that it could not straightforwardly infer how an engine would behave when struck by birds of different sizes from those on which it had been tested directly (FAA 2000).[9]

Another cause of contention lies in the fact that birds over 8 pounds are not considered in the tests at all, despite the wide variety of such birds that routinely encounter aircraft engines. Geese, storks, and swans all routinely reach greater weights—some exceeding 30 pounds—and, as noted previously, all three exist in large (and in some cases rapidly expanding) numbers. Flight 1549, which famously landed in the Hudson in 2009, was brought down by Canada geese, for example, as was a military Boeing 707 outside Anchorage in September 1995 (Marra et al. 2009).

The FAA contends that its tests adequately accommodate the risks posed by what it calls "very large" birds (i.e., those over 8 pounds) since the "large" bird test already assumes that the engine might be destroyed (defined as losing a single fan blade) and requires only that blade fragments be contained and the engine shut down safely. If the engine is destroyed safely by a large bird, the agency argues, it will also be destroyed by very large bird, and probably just as safely (BNME 1999; FAA 1998). However, this argument rests on two assumptions—that losing a single fan blade is the most catastrophic damage that a bird can cause; and that an engine is destroyed "safely" if the blade is contained by the engine housing—and critics point out that neither assumption is always borne out by experience. US Airways (1999), for example, has expressed doubt about both assumptions, citing a case where more than one blade was "liberated" after one of its aircraft ingested an Eider duck. The blades were contained as intended (which has not always been the case), but the strike seriously damaged the engine structure, almost breaking off the inlet cowl in a "potentially catastrophic" way.

SPECIES Other critics have concerns about the fact that the tests categorize birds by mass, but not by species. Aircraft in service ingest a wide variety of "avifauna"—from pelicans to parrots, 503 different bird species were involved in collisions with US aircraft between 1990 and 2013 (FAA 2014, vi; Langewiesche 2009a)—and it is easy to imagine that the strike effects of these birds might be determined by more than their mass alone. This fits with the discourse around bird strikes, which quite often highlights perceived distinctions between species. After an American Airlines flight ingested a cormorant in 2004, for instance, its spokesperson justified the large amount of damage by arguing that "a cormorant is chunkier, meatier and has more bones than a looser, watery bird [and] would have a harder time getting through the fan blades of the turbine" (Hilkevitch 2004).

The airline's assertion was certainly plausible. Different bird species have different shapes, consistencies, and densities, even when they have the same mass, and there are good reasons to believe that these variables significantly shape the birds' impact on engines (Budgey 2000). Take, for example, the aforementioned stipulation that a test bird (or birds) must strike the engine at its most vulnerable point for the test to be valid. To determine this point of vulnerability, engineers identify something called the "Critical Impact Parameter (CIP)." For most modern engines, this is usually the stress imparted to the leading edge of the fan blade (although other potential CIPs emphasize different engine elements, such as the blade root, and different variables, such as strain, deflection, or twist). Yet relatively small changes in the volume or density of the bird—such as those found among different species—affect the bird's "slice mass,"[10] which, in turn, can change the CIP and lead to a different test (FAA 2009, 3).

QUANTITY   A further debate surrounds the numbers of birds that the tests use. The volleys of eight "medium" and sixteen "small" birds, each fired in quick succession, are intended to simulate a flock encounter. Their similarity to actual flocks is limited, however, and some experts worry this makes the tests unrepresentative.

The most prominent argument made in this context is that the numbers are too small to adequately represent the most challenging conditions an engine might face. Critics point out that various birds congregate in dense groups, causing engines to sometimes ingest more of them than the tests anticipate. This is especially true of birds in the "small" weight category, such as starlings, which can gather in prodigious numbers. An MD-80 is reported to have left the snarge of 430 dead starlings on a runway at Dallas, for instance, and a Boeing 757 in Cincinnati is similarly said to have left over 400 (Eschen-felder 2000, 4). The fact that tests only require a single large bird raises similar questions, as these also sometimes flock. Geese, swans, and storks all gather in large numbers, for instance, especially around migration time.[11] An engine that gulps one goose, therefore, is reasonably likely to strike others, making for a more demanding challenge than that for which it was tested. When Flight 1549 was pulled from the Hudson, for example, the remains of multiple geese were found in both its engines (NTSB 2009, 80).

The fact that bird numbers might be higher than the tests suppose has complex ramifications for various aspects of bird-strike certification. The FAA treats multiple strikes to the same blade as an anomaly, for example, and has justified this rule by arguing that the tests use considerably *more* birds than are likely to be encountered in an actual bird strike (FAA 2004).

SPEED   Finally—for our purposes at least—there is a debate around the speed at which the birds should travel. As noted previously, certification guidance stipulates that they be launched at "takeoff speed," or about 200 knots, on the basis that the majority of bird strikes occur near ground level. Critics, however, point out that birds are still sometimes struck when aircraft are traveling higher, and therefore faster. (As of 2009, for example, the highest recorded bird strike was at 37,000 feet, with a griffon vulture off the coast of Africa [Croft 2009]). This has led some to suggest that 250 knots—the maximum airspeed allowed below 10,000 feet—would make a more challenging test (ALPA 1999). Others have pushed for even higher speeds.

Regulators have rebuffed these suggestions, claiming that bird-strike tests become less rather than more challenging when birds are traveling faster than 200 knots. The argument here again relates to the engine's Critical Impact Parameter. The FAA claims that the lower speed is more likely to "result in the highest bird slice mass absorbed by the blade at the worst impact angle, and therefore results in the highest blade stresses at the blade's critical location" (FAA 1998, 68641). Again, however, this is contested by many observers. The Airline Pilots Association (ALPA), for instance, questions whether the "slice mass" argument is a proven assumption. And the FAA itself has elsewhere concluded that strikes occurring above 500 feet—where airplanes are usually traveling faster than 200 knots—are more likely to cause damage than strikes at or below 500 feet (FAA 2014, xi).

MEANINGFUL DIFFERENCES

The misgivings outlined here are far from an exhaustive list of the concerns that experts have expressed about the representativeness of the FAA's bird-strike tests. Many other elements of the tests are contestable and contested. For example, there are debates about whether the explosions of air from the cannon affect the turbines and about whether the test engines should be attached to automatic recovery systems, as they would be under flight conditions (FAA 2001). Stepping back, we might also consider that the tests might be unrepresentative of the precision and care with which testers build and handle their test specimens: a dilemma common to many technological tests (Sims 1999, 501; MacKenzie 1989). The engines that experts use for certification testing are invariably new machines that have been pored over by scores of engineers, whereas those that encounter birds in real life will have seen some service and routine maintenance.

As should hopefully be clear by now, however, there are a potentially infinite number of ways in which bird-strike tests might imperfectly represent real bird strikes, and these potential discrepancies are important. The FAA has credible answers to many of the criticisms outlined here, but it has no *definitive* answers. The differences that critics have highlighted between the tests and real-world conditions might be irrelevant, but there are credible reasons to doubt this. Bird-strike experts can reasonably be confident that the colors of the birds and the day of the week are unlikely to affect their tests, but the issues outlined here represent meaningful uncertainties. Ultimately, there is no escaping the fact that the usefulness of bird-strike tests—what their results mean for an engine in service—hinges on a spectrum of contested relevance assumptions, concerning everything from the significance and accuracy of historical data, to the ways that engineers interpret that data and represent it in the laboratory.

The epistemological dilemmas that relevance assumptions pose for bird-strike tests do not stop here, however, for testers must also reckon with problems that arise from too *much* representativeness.

5.3 REPRODUCIBILITY

NARROWING OF VISION

Somewhat confoundingly, amid the concern surrounding the imperfect realism of bird-strike tests, there are critics who contend that the tests would be stronger if they were *less* realistic. Take, for instance, the following lament from a paper presented to the International Birdstrike Research Group:

> It has long been accepted that using real bird bodies in aircraft component testing is not ideal. The tests are not uniform. . . . Differences in bird body density between species and even between individuals of the same species may cause different and unpredictable effects upon impact, with consequent implications for testing standardizations throughout the world (Budgey 2000, 544).

Note how this argument invokes the variations among bird species outlined in this chapter, but it construes the problem that variation poses very differently: as something to be eliminated rather than faithfully reproduced.

This position is not as illogical as it might appear. All knowledge practices involve what Scott (1998, 11) describes as a "narrowing of vision," wherein experts seek to isolate key aspects of an otherwise "complex and unwieldy reality" by eliminating confounding variables. And STS scholars have long attested to the importance of such narrowing; arguing that it underpins every effort to classify, compare, and ultimately comprehend the world (e.g., Latour 1999, 24–79; Bowker and Star 2000). By this view, the usefulness of engineering tests stems, in part, from their ability to control the world so that engineers can isolate a single variable; which is to say their artificiality rather than their realism (e.g., Wynne 2003, 406; Henke 2000, 484; Porter 1994, 38). For instance, experts would struggle to assess the relative strengths of different metals if the tests of each metal were conducted at different temperatures or with samples of different purities.

Seen through this lens, it is important that bird-strike tests be reproducible. Partly because they would not be credible if they produced different results at different times: the same engine passing one day and failing the next. And partly because reproducibility offers insight: if every bird-strike test were unique, testers would struggle to compare the performances of different engines and would learn little about the factors that influence that performance. To be reproducible, however, the tests must be standardized; they must accurately represent each other, as well as the world itself. And achieving this means stripping away sources of variation—anything that "may cause different and unpredictable effects."

Hypothetically at least, it ought to be possible for engineers to maximize the reproducibility of most technological tests without compromising those tests' representativeness. They could do so simply by re-creating the variables that affect the phenomenon being tested, while controlling every-thing that does not: reducing the tests' realism (and thereby rendering them more legible and reproducible) without altering anything that matters to their outcome. This work would still force experts to grapple with the prob-lem of relevance, as they would have to identify every significant variable, but it would be mean they could pursue reproducibility and representative-ness at the same time.

When reliability is the specific variable being tested, as it is in bird-strike tests, however, then reproducibility is in direct conflict with representative-ness. The reason for this harks back to reliability's uniqueness as an engineer-ing variable. As outlined in chapter 2, reliability is inherently contextual: it is a measure of a system's failure performance in the real world and has no meaning independent of this context. This means that, unlike most engi-neering variables, it cannot be abstracted or isolated. In tests of reliability, therefore, reproducibility and representativeness inherently pull in opposite directions. The real world is inherently messy and complex, so any reduc-tion of that variability—any narrowing of vision—is, almost by definition, a potential reduction in the test's relevance. In these circumstances, testers must choose whether to prioritize reproducibility or representativeness, knowing that each comes at a cost to the other.

In the specific context of bird-strike tests, these tensions are well illus-trated by the debates that surround the use of artificial birds.

RUBBER CHICKENS

Engineers concerned with the reproducibility of bird-strike tests often advo-cate for the use of artificial birds. Usually made from gelatin and sometimes referred to as "cylinders of bird simulant material," artificial birds are the physical embodiment of an engineering ideal. This is to say that they are birds as seen through a lens that reveals only those variables deemed signifi-cant to ingestion tests. When Wile E. Coyote contemplates the Roadrunner in Warner Brothers' classic cartoons, he sees only dinner: roasted and aro-matic. Bird-strike engineers, in much the same vein, would see a potentially hazardous mass with a specific volume and uniform density. Neither is inter-ested in the subtleties.

Proponents of artificial birds often highlight advantages relating to their cost, convenience, and (usually obliquely) the intangible well-being and public relations convenience associated with firing fewer actual birds into

engines.[12] Above all else, however, they emphasize the virtues of reproducibility. With their gelatinous simplicity, artificial birds promise to reduce the
(literal and figurative) messiness of bird-strike tests: limiting the extraneous
variables involved in each strike to make those that count more susceptible
to measurement and calculation. They lack the feathers, bones, beaks, and
wings of a real bird, and much else besides; but, in this view, such features
are part of a complex and unwieldy reality that can be stripped away to
make birds more legible.

The benefits of reducing the realism of bird-strike tests in this fashion
undeniably come at a cost, however. Even proponents of artificial birds recognize that the projectiles do not accurately reproduce the complexities of
a real collision, and this makes the tests a less faithful representation of the
phenomenon they are intended to measure. This is to say that those proponents wish to remove certain variables, not because they are insignificant or
extraneous, but because they are the source of significant variations in the
data. Unsurprisingly, therefore, a lot of experts have misgivings about the
representativeness of artificial birds, many of which map onto the debates
about real birds outlined in this discussion, albeit usually in a more acute
form. (As outlined here, for example, some experts consider "toughness"
to be an important variable when modeling bird strikes, and yet artificial
birds are not designed to have the same toughness as a real bird [e.g., Edge
and Degrieck 1999]).

The debates around this are complicated by the fact that reproducibility
can sometimes be more useful than representativeness: assessment bureaucracies often preferring a test that ignores significant variables to one that
is difficult to repeat (Porter 1994). Indeed, I will argue in later chapters that
type certification leans heavily on comparing new systems with the systems
they are replacing, with greater reproducibility (facilitated by greater stan-
dardization) facilitates. (It being more useful for experts to explore the question *"how will this system perform in the world"* by approaching it through the
question *"how does this system perform relative to its predecessor,"* than by trying
to explore it directly.) Perhaps counterintuitively, therefore, the costs of artificiality might be worth paying.

For the time being, however, let us simply understand the problem of
reproducibility as a further epistemological complication: one more reason
to doubt that bird-strike tests can offer the kind of certainty that experts
impute to them.

5.4   AN ENDEMIC DILEMMA

"NOT ALL THE WAY"

Bird-strike tests are a modest fraction of certification's test regimen, but they illustrate a much more generalizable principle. Far from being exceptional, their ambiguities and uncertainties speak to an indelible gap between every certification test and the phenomena that those tests aspire to represent. (Later chapters will outline many similar debates and controversies around other systems and variables.) "Tests get engineers closer to the real world," as Pinch (1993, 26) puts it, "but not all the way." The relevance questions that permeate them are never solved. With a potentially infinite number of variables that might be significant in any test, experts cannot be certain they have identified them all or even that they have adequately represented the variables they identified.

Recognizing these uncertainties of representation, and the problem of relevance that they imply, is not the same as claiming that technological tests are irrelevant, valueless, or open to any interpretation.[13] Per chapter 2, however, it is important to understand that these uncertainties matter in contexts where experts are looking to establish extreme reliability, far more than they matter in other engineering contexts. Claims about reliability are essentially claims about *confidence*, and a small but crucial element of any claim about jetliner reliability is an extreme confidence in their engines' ability to ingest birds safely. That level of confidence is incompatible with the irreducible uncertainties of real bird-strike tests.

For engineers to properly grapple with ultrahigh reliabilities via tests they would need their tests to represent even the most esoteric variables: the kinds of obscure conditions and interactions that appear only once in every billion flights. Any expert claiming to have established a system's ultrahigh reliability in a test, therefore, would be asserting (with extreme certainty) that they had accurately reproduced all the most confounding and unusual conditions that system will face in the entirety of its messy operational life (those that occur only with "a specific phase of the Moon or something," as one correspondent put it). This is not realistic.

This is all to say that testing alone is a woefully insufficient explanation for the performance of modern aircraft, and for the FAA's ability to accurately predict that performance. A close examination of its real practices amply illustrates the finitist case against ultrahigh reliability management.

Still, however, the manifest insufficiency of testing in this regard does not diminish the empirical facts of aviation safety, which is as evident in the bird-strike data as anywhere. Catastrophic bird-strike incidents like Flight 1549 do occur, but there is no statistical basis to claim that engines are less bird resilient than experts claim. The actuarial data speak louder than any lab test: birds very rarely bring down commercial jetliners. If anything, the resilience of engines to birds appears to be increasing. The percentage of wildlife strikes that cause damage has declined from 20 percent in 1990 to 5 percent in 2013 (FAA 2014, xi). In this, as in countless other areas, therefore, aviation experts are transcending the limits of their tests.

The aviation paradox remains.

For a better answer to that paradox—a means by which experts could potentially be transcending the limitations of their tests—we might look to systems-level engineering, whereby engineers compensate for uncertainties in their designs (and their tests of those designs) by making them failure tolerant. It is to this that we now turn.

6 THE SUM OF ALL PARTS: MODELING RELIABILITY WITH REDUNDANCY

It is well to moor your bark with two anchors.
—Publilius Syrus

6.1 SYSTEM-LEVEL SAFETY

"IMPOSSIBLE" FAILURES

On June 24, 1982, 37,000 feet over the Indian Ocean, the pilots of British Airways Flight 009 noticed the power begin to drop on each of the Boeing 747's four engines. Their concerns were soon heightened when pungent smoke began creeping into the flight deck. And they intensified even further when passengers reported seeing a numinous blue glow around the engine housings. Just minutes after the first indication of trouble, one engine shut down completely. Soon a second failed; then a third; and then, upsettingly, the fourth. With no understanding of why, the pilots found themselves gliding a powerless jetliner, 80 miles from land and 180 miles from a viable runway.

Despite a memorably laconic address from the captain,[1] the glowing, silent engines and creeping smoke understandably caused some consternation among the 248 passengers. Several scribbled heartfelt notes to loved ones as their airplane gradually earned a mention in the *Guinness Book of Records* for "Longest unpowered flight by a non-purpose-built aircraft."[2] Just as cabin crew were preparing them for an unpromising midocean ditching, however, the power abruptly returned. Passengers in life jackets applauded the euphony of restarting engines, and—after some last-minute drama with an occluded windscreen[3]—the pilots deposited them safely in Jakarta (Tootell 1985).

The underlying cause of this aeronautical scare—now widely remembered as the "Jakarta incident"—was not a mystery for long. Investigators found the engines were full of ash particles and quickly deduced that the airplane had inadvertently flown through the plume of a volcano erupting nearby, the ash from which had clogged the engines and sandblasted the cockpit windows. The last-minute reprieve had come when the plane lost enough altitude to leave the plume, whereupon airflow had cleared some of the ash.

Although easily explained, the Jakarta incident unnerved the aviation community, which was (and largely remains) committed to the belief that no commercial jetliner, especially a four-engine 747, could lose propulsion from all its engines simultaneously. The 747 can fly relatively safely on a single engine, so four give it a comfortable margin of security. Under normal circumstances, experts consider even a single engine failure to be highly unlikely. So until British Airways inadvertently demonstrated otherwise, they considered the likelihood of all four failing simultaneously to be functionally zero: as near to impossible as makes no practical difference. The industry was so confident of this that pilots were trained to treat indications of quadruple engine failure as evidence that there was something wrong with the indicators, not the engines. Nobody had considered the possibility of volcanic ash.

Flight 009's near-miss was not the first time that the industry found that it had placed too much confidence in redundancy, however, and it would not be the last. Galison (2000), for example, discusses a similarly unanticipated failure in a redundant aviation system, this time with an unhappier ending. United Airlines Flight 232, a DC-10 en route to Chicago in 1989, lost its hydraulic systems after its tail-mounted engine exploded. McDonnell Douglas had explicitly designed the DC-10's hydraulics to resist such trauma. Each channel was triple-redundant, with redundant pumps, redundant reservoirs of hydraulic fluid, and redundant (and differently designed) power sources. And this redundancy had led the aviation community to deem a total hydraulic failure in the DC-10 to be functionally impossible, just as it had done with engine failure. This was "so straightforward and readily obvious," experts had argued, that "any knowledgeable, experienced person would unequivocally conclude that the failure mode would not occur" (Haynes 1991). Nevertheless, the engine explosion managed to sever all three systems, rendering the airplane all but uncontrollable. (Remarkably, there was a credible attempt at a crash landing. The pilots, together with a flight instructor who happened to be on board, somehow managed to steer the plane to a runway by manipulating the power to different engines. This feat—which nobody was subsequently able to replicate in the simulator—saved the lives of 185 of the 296 passengers and crew.)

These incidents, and others like them, are significant because the reliability of all civil aircraft, and all assessments of that reliability, depend heavily on notions of redundancy. It would be fair to say, in fact, that redundancy is a sine qua non of almost all catastrophic-technological design and assessment.

DEFINITIONS AND AMBIGUITIES

Much like reliability itself, "redundancy" is a seemingly intuitive concept that is nevertheless surprisingly, and perhaps revealingly, difficult to pin to the page. As with reliability, the term has a specific meaning in engineering that is both similar to and sometimes at odds with its common English usage. In both contexts, it implies some manner of repeating, but where it usually has negative connotations in its common usage—as something superfluous and excessive—engineers more often equate it with surety. In the latter context, it is understood to be a property of a system's architecture. Broadly, if we imagine a "system" as being composed of many "elements," then an element can be described as being "redundant" insofar as it is part of a system that contains backups to do that element's work if it fails; and a system can be described as "having redundancy" insofar as it consists of redundant elements.

So far, so simple, but note that the definitions given here are inherently contextual. They can operate at multiple levels, such that, for instance, redundancy can take the form of multiple capacitors on a circuit board; multiple circuit boards on a sensor; multiple sensors in a system; and so on, all the way to the duplication of entire sociotechnical networks. (In the US missile programs of the 1950s, for example, the Air Force entrusted mission success to the doctrine of "overkill" and the redundancy of the whole missile and launch apparatus [Swenson, Grimwood, and Alexander 1998, 182]). A further ambiguity in such frameworks is that some redundant elements, such as engines on an airplane, work concurrently (but are capable of carrying the load by themselves if required), whereas others, such as backup generators, usually lie idle, only to awake if and when they are needed.

Jetliners, like most catastrophic technologies, invoke different forms of redundancy on many levels simultaneously. Indeed, the type-certification

process requires them to do so. Redundancy has been a staple design requirement for airframes since the dawn of aviation regulation. (The very first
airworthiness standards, for instance, required that biplanes have duplicate
flight control cables [FAA 2002c, 23–24]). Its significance to aeronautical
engineering and oversight is difficult to overstate. A core tenet of the FAA's
safety-philosophy is that no single component-level failure should be catastrophic. Achieving this goal, and formally demonstrating that achievement,
would both be unthinkable without redundancy.[4]

RELIABILITY SQUARED

Perrow (1984, 196) equates redundancy with a Kuhnian paradigm, and while
invocations of Thomas Kuhn are often hollow, the analogy is uncommonly
useful in this context. It highlights the way that redundancy acts as a conceptual lens through which experts "know" artifacts: shaping designs by shaping
how those designs are understood.[5]

Redundancy's value to engineers of catastrophic technologies lies in two
closely related, but nevertheless distinct, reliability-related functions that it
performs. The first is that it allows experts to *design* complex systems that are
far more reliable than their constitutive elements. (The realization that complex systems could be made more dependable than their components came
surprisingly late, for reasons that I will discuss, but it was a breakthrough
nonetheless.) The second relates to what the FAA calls "checkability," its
slightly unfortunate term for a system's capacity to be assessed and verified.
It is that redundancy allows experts to quantitatively *demonstrate* (as well as
achieve) higher levels of reliability than would be possible with tests alone.

To understand why this second function is important, it is useful to
understand that even if (contra the previous chapter) we imagined tests to
be perfectly representative, they alone would still be unsuitable for demonstrating the extreme reliability required of catastrophic technologies.
The precise amount of test data required to statistically establish a defined
level of reliability is contested at the margins (e.g., Ahmed and Chateauneuf
2011; Littlewood and Wright 1997), but at the levels required of catastrophic
technologies it is unambiguously impractical. Rushby (1993), for instance,
calculates that a single test system would need to run failure free for over
114,000 years to demonstrate a mean-time-to-failure in excess of a billion
hours. Such times could be reduced by running multiple tests in parallel,
but the cost of reducing them enough this way would be gigantic. Even by

certification's own logic, untroubled by the problem of relevance, therefore, jetliner systems would be insufficiently "checkable" if tests were the only means of establishing reliability.

Redundancy offers a solution to this checkability problem. It can serve this second function, in essence, because the way it performs the first (i.e., the manner in which it increases the reliability of a system) can be represented mathematically. Put very simply: where two redundant elements operate in parallel to form a system, then the probability of that system failing can be expressed as the probability of one element failing multiplied by that of the other failing. If tests show that one element has a 0.0001 probability of failing over a given period, for instance, then the probability of two identical elements failing over the same period would be that number squared—that is, $(0.0001)^2$ or 0.00000001. Note that this calculation has demonstrated a *ten-thousandfold* increase in reliability, and has done so without any recourse to more lab tests (Shinners 1967, 56; Littlewood, Popov, and Strigini 2002, 781; FAA 1982, 3).

It is this capacity for transcending tests that makes redundancy so valuable to catastrophic technology assessment processes like type certification. Although not the only systems-level tool for *increasing* a system's reliability—some elements can be made fail-safe instead, for instance—the manner in which redundancy can be modeled in calculations makes it almost indispensable to the processes by which experts formally *demonstrate* ultrahigh levels of reliability (Downer 2011a; 2009b).

For all their importance and utility, however, redundancy calculations, much like bird-strike tests, are far messier in practice than they appear on paper. This is because models of systems, much like tests of systems, are representations, the fidelity of which raises unanswerable questions. And while redundancy's relationship with reliability might seem straightforward when represented mathematically, its practical implementation involves reckoning with complex uncertainties and judgments that are difficult to capture with quantitative certainty.

6.2 MESSY PRACTICE

IMPERFECT REPRESENTATIONS

Redundancy is undoubtedly a powerful engineering tool. It has allowed engineers to make almost all complex systems—certainly all catastrophic technologies—more reliable than their constituent elements. Despite its benefits, however, evidence shows that redundant systems are never quite as reliable as formal models would imply (Littlewood 1996). The causes of this deficit are readily apparent when the discourse around redundancy is examined closely, as any such examination reveals routine disagreements about the appropriateness of various assumptions and variables. Such disagreements take many forms, but the following sections sketch out some of their more prominent themes. By outlining issues pertaining to "mediation," "common-cause failures," and "failure propagation," in turn, I will unpack some of redundancy's practical complexities, and show how those complexities give rise to uncertainties that are not easily reduced to definitive numbers.

Let us turn first to redundancy's relationship with mediation.

MEDIATION   John von Neumann, the Hungarian émigré polymath, is generally credited as being the first person to propose redundancy as a way of increasing the reliability of complex, tightly integrated technological systems, in a 1956 treatise: "Probabilistic Logics and Synthesis of Reliable Organisms from Unreliable Components." The mid-1950s might seem incongruously late for such a straightforward-seeming insight to gain traction, but the true innovation lay not in the idea of redundancy per se, so much as in envisaging a system that could invoke it without human intervention. Grappling with the headache of aggregating thousands of unfaithful vacuum tubes into a working computer, von Neumann intuited that redundancy could help if it were combined with a managerial system to immediately identify and mediate between failures: recognizing malfunctioning tubes and switching to their backups without interrupting the wider system. It was this management system—and the problem of automatically distinguishing failed from functional tubes—that posed the real engineering challenge.

To resolve the mediation problem, von Neumann proposed a "voting" system. "The basic idea . . . is very simple," he wrote, "Instead of running the incoming data into a single machine, the same information is simultaneously fed into a number of identical machines, and the result that comes out of a majority of these machines is assumed to be true. . . . this technique can be used to control error" (1956, 44). Not every redundant element in a system requires mediation in this fashion—redundant load paths in airframe fuselages don't, for instance—but most do. And by devising a way to automate mediation, von Neumann made redundancy a staple tool of high-reliability engineering.

His solution was no panacea, however, as mediating systems pose their own costs, challenges, and uncertainties.

One often-underappreciated cost of mediation is that it makes systems more complex. Even before mediation, redundancy increases the number of elements in a system, making unexpected interactions more likely and fail-ure behavior more difficult to verify. (So it is that scholars routinely correlate simplicity with reliability in technological systems [e.g., Perrow 1984, 270; Kaldor 1981, 111; Arthur 2009]). This can be a problem even with unmedi-ated redundancy, but it is greatly exacerbated by adding systems to mediate between elements. So much so, in fact, that experts sometimes argue that mediated redundancy can increase the complexity of a system to a point where it becomes a primary source of *unreliability* (Rushby 1993; Hopkins 1999).

Take, for example, the engines on a jetliner. It may sound simple to affix more engines to an airplane, but this simplicity quickly dissolves if we con-sider the elaborate, highly integrated management systems needed to govern them. Failures in redundant jetliner engines instigate an orchestra of rapid and highly automated management interventions. A computer first deter-mines which engine has failed, for instance, and then it indicates the failure to the pilot, cuts the fuel, douses any flames, adjusts the rudder, compensates for the missing thrust, and much more (Rozell 1996). Many of these actions are safety critical—especially during takeoff, when reaction margins are tight and even a momentary loss of power can be fatal—and all require a suite of sensors and computers with high-authority connections to disparate ele-ments of the aircraft. All this inevitably makes the system's failure behavior more difficult to understand and control. It also creates entirely new avenues of failure; a faulty computer that erroneously shut down engines during take-off could easily fell a jetliner.

This example also illustrates a second dilemma of mediation: that man-agement systems are themselves critical systems that require extreme reli-ability. Air-accident investigations frequently implicate mediating systems as contributory factors to catastrophic failures. When USAir Flight 427 crashed while approaching Pittsburgh in 1994, for instance, investigators found that the Boeing 737's computer had failed to adequately compensate for the roll generated by a rudder failure (NTSB 2006b, 7). Mediating systems have even been known to instigate aviation disasters. After Indonesia AirAsia Flight 8501 crashed into the Java Sea in 2014, for instance, investigators identified a failure in the Airbus A320's failure-monitoring systems as a primary cause. They found that a malfunctioning sensor in the rudder had led the crew to reset the computer, which disabled the autopilot, which in turn led them to stall the airplane (BBC 2015; KNKT 2015).[6]

The criticality of mediating systems raises especially challenging questions for the designers and regulators of catastrophic technologies because such systems cannot themselves be redundant. This is because the mediating elements themselves would then need mediating, and so on, in an infinite regress. "The daunting truth," to quote a 1993 report to the FAA, "is that some of the core [mediating] mechanisms in fault-tolerant systems are single points of failure: they just have to work correctly" (Rushby 1993). Engineers usually negotiate this problem by designing mediating systems to be simpler, and thus (hopefully) more reliable, than the systems they mediate. But while this arguably makes the problem tractable from a design perspective, it does little for experts hoping to use abstract models of redundancy to demonstrate ultrahigh levels of reliability.

Experts manage this latter problem—of integrating mediation into models of ultrahigh reliability—in different ways. One is with semantics: certification standards often just don't define mediating systems as "safety critical," so the reliability of those systems is not held to the same standard of proof. Another is by giving mediating functions to human beings. Type certification is an analysis of the airplane, largely unsullied by the vagaries of its operation. (The FAA does assess pilots, but via a separate process; certification essentially treats them as a solved problem.)[7] If pilots are made responsible for mediating between elements, therefore, assessors can effectively "lose" any uncertainties arising from mediation by exploiting the interstices between different regulatory regimes.

The role that humans often play in this regard make them uniquely important to understanding redundancy mediation, and it is worth pausing to examine this role in slightly more depth. From a design perspective, the fact that people sometimes can respond creatively to unanticipated errors and interactions can make them uniquely versatile as mediators between redundant elements (Hollnagel, Woods, and Leveson 2006, 4; Rasmussen 1983). Recall, for instance, the crew of Flight 232, outlined previously, who used the engine throttles to steer a DC-10 after its hydraulics failed. Again, however, delegating mediating tasks to humans is far from a perfect design solution, and it does little to resolve the problems that mediation poses to reliability calculations (Bainbridge 1983).

There are two basic reasons for this. The first is that relying on human beings rarely negates the need for safety-critical mediating elements. This is because humans can rarely mediate effectively without relying on an elaborate series of indicators and sensors working reliably. (As in the example of Flight 8501, for instance, it is not uncommon for indicator failures to be implicated in accidents.) The second, more fundamental issue is that people are proverbially imperfect. They get ill; they get confused; they run a gamut of emotions from stress to boredom; and for all these reasons, they sometimes make mistakes, disobey rules,[8] and, very occasionally, commit premeditated acts of sabotage (Reason 1990; Perrow 1999, 144–146; Lauber 1989). It would be fair to say, in fact, that human beings are a significant source of failure in jetliners, as they are in all sociotechnical systems (Dumas 1999; Reason 1990; Bainbridge 1983). The NTSB has estimated that 43 percent of fatal accidents involving commercial jetliners are initiated by pilot error (Lewis 1990, 196).

Airframers work hard to mitigate the risks posed by human mediation (many of which might reasonably be attributed to ergonomic factors, such as misleading cockpit displays [Perrow 1983; Rasmussen 1990]). They hone the designs of human-machine interfaces with an eye to making them intuitive and error tolerant ("foolproof," or sometimes "drool-proof" in less reverent industries). They also limit the scope of pilots' actions with elaborate flight protections, which, when enabled, sometimes allow the computer to overrule commands it deems dangerous. They even invoke further redundancy, designing jetliners to be crewed by two pilots.

Such measures are far from perfect, however, and sometimes create epiphenomenal dilemmas (Bainbridge 1983). Redundant personnel can induce overconfidence, for example, or what Sagan (2004, 939–941) calls "social shirking": a mutual belief that the other person will "take up any slack." (This can threaten communications between pilots, for instance [e.g., Wiener et al. 1993]). Automated flight protections can have a similar effect, as in 1988, when the captain of an Airbus A320 wrongly assumed its flight computer would prevent a stall during a low-speed airshow flyover (Macarthur and Tesch 1999). (This was the A320's first passenger flight. Thousands of spectators watched as the airplane, full of raffle winners and journalists, crashed into a forest at the end of airfield.) There are also concerns that such protections undermine the basic competencies of pilots, who are decreasingly required to exercise even rudimentary flight skills outside the simulator. (The 2009 loss of Air France Flight 447, for instance, is often partly attributed to a basic pilot error, which many observers attribute to the universality of automatic flight protections [Langewiesche 2014]).

COMMON CAUSE A second dilemma of redundancy revolves around the independence of redundant elements. Recall Flight 009, with its dramatic power loss outside Jakarta, and Flight 1549, with its remarkable landing in the Hudson River. Both were precipitated by multiple, simultaneous engine failures caused by a common external pressure: volcanic ash in the former case, and geese in the latter. Engineers refer to such incidents—where redundant elements fail at the same time for the same reason—as "common mode failures," and the FAA (1982, appx. 1) has described them as a "persistent problem" for certification assessments. The accidents that they cause are significant in this context because they highlight a prevalent critique of redundancy calculations: that they often erroneously assume that redundant elements will behave independently with respect to their failure behavior (e.g., Popov et al. 2003).

The elegant mathematics outlined previously—where the reliability of one redundant element can be multiplied by that of another to arrive at the combined reliability of the system—only works insofar as those reliabilities are assumed to be perfectly independent, which is to say that the chances of one element failing are not linked, in any way, to the chances of the other failing. As the 155 passengers who began their flight at LaGuardia and ended it in the river can attest, however, independence is rarely perfect.

There are many principled reasons to doubt the independence of redundant elements in a system. Almost by definition, for instance, redundant elements share a common function. Many are also colocated, share a common design, and draw on a shared resource, such as fuel or electricity. Such commonalities inevitably create common vulnerabilities. They mean that a cloud of ash or a flock of birds is likely to stress all the engines on a jetliner simultaneously, for instance, as might a fuel leak or even an imperfect maintenance operation (Eckhard and Lee 1985; Hughes 1987; Littlewood 1996; Littlewood and Miller 1989).[9] Where redundant elements operate at the same time (as engines do), moreover, then they are likely to fatigue in similar ways, further aligning their failure behavior. (Although it is also entirely possible for external pressures to stress even idle elements as they wait in reserve [Acohido 1996].)[10]

Just as engineers have practical design techniques for managing the complications of mediation, so they have ways of maximizing the independence of redundant elements. Most rely on what is known as "design diversity": the practice of designing redundant elements differently while keeping their functions the same. The idea is to create elements with dissimilar weaknesses, and thus more independent failure behavior. Manufacturers pursue diversity in varying ways. Some leave it to evolve spontaneously, contracting isolated different engineering teams (often from separate contractors) to design redundant elements and trusting that a lack of central authority will create sufficient variation (Littlewood and Strigini 1993, 9; Bishop 1995). The A320's redundant flight computers are supplied by different vendors for this reason, for example, as is the software they run (Beatson 1989). Others actively try to force diversity by explicitly requiring different teams to use divergent approaches, solutions, and testing regimens. Software manufacturers, for instance, sometimes require teams to program in different languages (Popov et al. 2003, 346). An elaboration of this approach, known as "functional diversity," requires that engineers design elements to use different inputs, in the hope that conditions that challenge one will not challenge another (Littlewood, Popov, and Strigini 1999, 2; Beatson 1989). So it is, for instance, that jetliners simultaneously use pressure, radar, and global positioning systems (GPS) to determine altitude. (The same principle is at work when pilots on the same flight are required to eat different meals to protect against simultaneous food poisoning.)

All these strategies can be useful for improving the reliability of a system, but, again, there are many principled reasons to believe they can never be perfect. The idea that different groups, when left to their own devices, will design the same system differently is bolstered by studies highlighting the contingency of technological designs (e.g., Bijker, Hughes, and Pinch 1989), for example, but the same research suggests that their ideas will likely converge when designers come from similar professional cultures and have problems specified in similar ways. (It has been found, for example, that programmers asked to independently design different versions of the same software tend to make similar mistakes and produce code with coincident failure behaviors [Knight and Leveson 1986]).

"Forced diversity" might be stronger, but it too has theoretical shortcomings. It demands a well-defined notion of "dissimilarity," which in turn poses questions about what constitutes a "meaningful difference." But "difference," much like "representativeness," always has a bounded and socially negotiated meaning (Collins 1985, 65). Like truth, beauty, and contact lenses, it inevitably rests in the eye of the beholder and must be restricted to a finite number of variables before engineers can enforce it on designs. (Should diverse elements be forced to use wires of different gauges, for example? Should they both use wires? Should they both use electricity?)

"Functional diversity" shares the same underlying problem and has its own distinctive limitations. Designing systems to operate on different inputs might help in some circumstances, but the exogenous pressures on a system often come from sources that are unrelated to its inputs, such as a fire, collision or lightning strike. It is also true that seemingly different and separate inputs are often interrelated; extremes of temperature, for instance, correlating (at least loosely) with extremes in pressure. And also that, even when building functionally dissimilar systems, engineers are usually still working from a similar definition of what constitutes a "normal" or "routine" environment (Littlewood and Strigini 1993, 10).

For all these reasons, it would be misleading to imagine that design diversity, in whatever form, produces mathematically perfect failure independence. (As with mediation, it has even been the *source* of failures. The first Space Shuttle launch, for instance, was initially scrubbed after its primary and backup software showed a 40-millisecond time skew. The problem was caused by different programming priorities chosen by their— purposefully different—programmers: IBM and Rockwell [White 2016, 350– 351]). To the extent that diversity does provide additional independence, moreover, it would be misleading to imagine that experts could precisely quantify the increase. To do so would require testing elements together as a single system to measure the rate of coincident failures: an endeavor that would quickly run into the practical limits that experts use redundancy to transcend. The precise degree of independence between elements is inherently uncertain, therefore, and techniques like diversity do little for those who would use redundancy to quantify (as opposed to maximize) the reliability of a critical system.

PROPAGATION   A third complication of redundancy arises from the fact that failures in energetic and interdependent systems rarely keep to themselves. An important dimension of technical malfunction that redundancy models struggle to capture is that failures have a tendency to propagate across elements in what engineers call "cascades" (NAS 1980, 41; Zdzislaw, Szczepanski, and Balazinski 2007).

Consider, for example, a 2010 incident onboard Qantas Flight 32: a two-decked Airbus A380 carrying over 450 passengers. The airplane's pilots were taken aback when their cockpit lit up with fifty-four simultaneous failure indications. (The cockpit designers had never envisaged so many coincident malfunctions, which filled its displays with more text than they were able to show [Lowy 2010]). After a dramatic overspeed landing in Singapore—involving an emergency gear drop, four blown tires, and a three-hour dousing by emergency crews—it was discovered that the many-pronged crisis had been instigated by a single misaligned counterbore in one of the airplane's oil pipes. The misalignment had led to a fatigue fracture; the fractured oil pipe had led to an engine fire; the fire had caused one of the engine turbines to shatter, releasing three disk fragments through the engine housing; and the disk fragments had wreaked havoc on the airframe. They ripped through a wing, damaging a structural element and slicing electrical cables; they punctured two fuel tanks, causing leaks and more fires; they disabled the airplane's antilock brakes, one of its hydraulic systems, and the controls for one of the other engines; and they damaged the landing flaps. These many insults, in turn, instigated a range of other downstream failures—for instance, the leaking fuel created a mass imbalance that pilots were unable to redress because of electrical damage to the pumps—which, together with the primary failures, created the unmanageable kaleidoscope of cockpit warnings (ATSB 2013; Lowy 2010).

On one level, the fact that failures tend to instigate more failures is a straightforward observation, but it has complex ramifications for reliability calculations. This is because it means that a complete understanding of the reliability offered by redundancy needs to account, not only for the independence of redundant elements in a system, but also for the independence of these elements from other, functionally unrelated elements in other systems. Accounting for this kind of failure propagation can be extremely challenging, however, because its mechanisms are often difficult to predict, measure, and control.

There are many reasons why failure propagation across systems is difficult to model. One is that it is difficult to always predict *how* elements will fail. "You can't always be sure your toilet paper is going to tear along the perforated line," as one aviation engineer put it.[11] In 2005, for instance, Malaysia Airlines Flight 124—a Boeing 777 taking off from Perth—spontaneously pitched upward, activating stall warnings and startling the crew. On its return to the airport, investigators identified the cause to be a faulty accelerometer. The accelerometer had had a redundant backup in case it failed, but its designers had failed to predict the *way* that it would fail. They had assumed that a failure would always result in an output of zero volts, but in this instance it had produced a high-voltage output that confused the flight computer (ATSB 2007).

Beyond this, however, even very predictable failure mechanisms can introduce difficult uncertainties if they propagate in ways that are difficult to contain. Engineers are well aware that explosions are plausible failure modes in some elements, for example, and that explosions can jeopardize other, functionally unrelated elements, but they are hard pressed to mitigate such effects. (Several commercial aircraft have been lost after their fuel tanks exploded. TWA Flight 800, which blew up in 1996 for undetermined reasons shortly after leaving JFK Airport, is one. Pan Am Flight 214, which blew up after being struck by lightning in 1963, is another.) And, while explosions are exemplary forms of propagation, there are many other ways in which the failure of one element can have unexpected consequences throughout an airframe. A failed element might threaten others simply by virtue of having a destabilizing mass, for instance, or by drawing excessively on a common resource. (This happened in August 2001 when a faulty crossfeed valve near the right engine of an Airbus A330-200 bled fuel until none remained to power the aircraft.)[12]

Accounting for propagation can have counterintuitive implications for redundancy's relationship to reliability. Consider, for example, the safety afforded to an airplane by redundant engines. "Two engines are better than one," writes Perrow (1984, 128), and "four better than two." This seems simple enough, and perhaps Perrow is correct, but his truism is more questionable than it appears. If we recognize that when engines fail, they can do so in ways that propagate and catastrophically damage the airframe—as happened to Qantas Flight 32, for instance, or, more fatally, to United Flight 232 in

July 1989—then it is less clear that four engines are safer than two. Indeed, it is possible that four engines could be significantly *less* safe than two.

To understand how two engines could be safer than four, imagine a hypothetical airplane that can function safely with one engine and is sold in both a two- and a four-engine configuration. For the sake of illustration, let us assume these engines have perfectly independent failure behavior. Now let us imagine that the chances of any given engine failing during a flight are one in ten. (It is an extraordinarily unreliable design!) To make matters worse, however, one out of every five engines that fail also explodes, destroying the airplane. (Of course, the airplane is also lost if all the engines fail in the same flight.) In this scenario, the two-engine aircraft would have a higher chance of an "all-engine" failure, but a lower chance of experiencing a catastrophic explosion. The math works out such that the combined risk of any catastrophic event during flight (an all-engine failure or an explosion) is higher with a four-engine configuration than with two. This is to say that two engines would be safer than four.

So it is that propagation might theoretically create circumstances where added redundancy *detracts* from a system's reliability. This is more than simply academic. The engine example given here uses unrealistic probabilities but it captures a real concern. Boeing has made essentially the same argument about two- versus four-engine aircraft, albeit with less straightforward numbers. Advocating for fewer restrictions on two-engine airplanes, the company claimed that its 777 was safer with two engines because of the reduced risk of one failing catastrophically (Sagan 2004, 938).

As with the other complications of redundancy outlined here, engineers have pragmatic design techniques for mitigating propagation. These often involve "isolating" different elements by physically segregating and/or shielding them from each other. Critical microelectronics are sometimes encased in ceramic, for example, while engines are separated on the wing with cowlings designed to contain shrapnel from broken fan blades.

Much like diversity, however, and for essentially the same reasons, isolation is a necessarily interpretive, ambiguous, and unquantifiable property of systems. As with the measures that engineers use to mitigate redundancy's other complications, therefore, it is a useful but imperfect practice that gives rise to complex disagreements. During the certification of the Boeing 747-400, for example, the FAA and its European counterpart differently interpreted an identically worded stipulation governing the isolation of redundant wiring (FAA 2002b, 4–6). The European regulator interpreted the word "segregation" more conservatively than the Americans, forcing Boeing to redesign the wiring late in the certification process. Because of this, two different designs of the 747-400 coexisted (GAO 1992, 16).

Engineers charged with certifying aircraft assess isolation with analytical tools such as Failure Modes and Effect Analysis, which maps the relationships between elements. Applying these tools is as much art as science, however, it being impossible to foresee every possible interaction (and to calculate the limits of one's forsight), and they are more useful for design than for assessment (NTSB 2006b; Hollnagel 2006). As with the ambiguities arising from mediation and common cause failures, therefore, the type-certification process ultimately navigates the difficulties of quantifying isolation by fiat. It mandates certain deterministic requirements—such as engine separation—and then unrealistically treats those requirements as offering mathematically perfect independence (FAA 2002b; NAS 1980; Leveson et al. 2009).[13]

6.3 OBDURATE UNCERTAINTY

AN IMPERFECT TOOL

As in chapter 5's discussion of bird-strike tests, none of the complications outlined here should be read as a suggestion that redundancy is not a useful engineering tool for reliability. (Although they do suggest that it isn't always the *optimal* design solution for reliability. As Popov et al. [2003, 346] put it: "Redundancy [is only] a reasonable use of project resources if it delivers a substantial . . . increase in reliability, greater than would be delivered by spending the same amount on other ways of improving reliability.") For our purposes, however, the important insight to be gleaned from this discussion is that redundancy is an *imperfect* tool—one that requires subjective judgments and complex interpretations to implement.

These judgments and interpretations are important. The uncertainties they imply are all potential sources of error, where every error holds the potential for catastrophe. The FAA has conceded as much, noting in 2002 that redundancy "has costs, complexities, and the inherent risk of unforeseen failure conditions associated with it" (FAA 2002c, 23–24). Its type-certification processes are framed around quantitative reliability targets, however, and models of redundancy are all but indispensable for demonstrating compliance with those targets. So experts elide redundancy's uncertainties in their reliability calculations, thereby making certification possible but doing little to negate the dangers that those uncertainties pose (which, even if marginal, ought to be intolerable in contexts requiring ultrahigh reliabilities.)

As with bird-strike tests, therefore, a close look at the nuances of redundancy modeling does more to illustrate the aviation paradox than to resolve it. Type certification is supposed to be the process through which experts establish the ultrahigh reliability of jetliners, but, epistemologically, its imperfect tools and practices are simply not up to the job. Examined closely, the practical limitations of tests and models are impossible to reconcile with the service record of civil aviation. The ultrahigh reliability of modern jetliners, as well as the fact that regulators accurately predict that reliability prior to them entering service, imply a depth of understanding that is incommensurable with the practices from which that understanding is ostensibly derived. Logically, at least, jetliners should be failing for reasons that slip though even the most rigorous analyses.

And some do exactly that.

7 RATIONAL ACCIDENTS: ON FINITISM'S CATASTROPHIC IMPLICATIONS

The best laid plans o' mice an' men / Gang aft agly.
—Robert Burns

7.1 ERROR PLANE

737-CABRIOLET

On April 28, 1988, Aloha Airlines Flight 243, a passenger-laden Boeing 737, left Hilo Airport on a short hop between Hawaiian islands. It climbed gently to the trip's cruising altitude of 24,000 feet. And then it tore apart.

The airplane's pilots would later report a savage lurch, followed by a tremendous "whoosh" of air that tore the cabin door from its hinges. One recalled glancing back through the space where the door used to be and seeing blue sky instead of the first-class cabin (NTSB 1989, 2). Closer inspection, had there been time, would have revealed passengers silhouetted against the emptiness, still strapped to their seats but no longer surrounded by an airplane fuselage. All of them hurtling through the air, unshielded, at hundreds of miles per hour; far above the open ocean.

Unable to communicate over the howling winds, but finding they were still in control of the airplane, the pilots set the emergency transponder and landed gingerly at nearby Kahului. Once safely on the tarmac, the airplane's condition indelibly marked its place in the annals of aeronautics folklore. An eighteen-foot, hemispherical fuselage section—thirty-five square meters of the first-class cabin—had ripped away from the airframe, severing major

FIGURE 7.1
Aloha Airlines Flight 243. *Source:* Hawaii State Archives.

structural beams and important control cables. In the affected section, only the floor and seats remained intact (figure 7.1).

The human toll was less than could reasonably have been hoped. Sixty-five of the ninety passengers and crew had been injured by winds and flying debris, eight seriously. But, by grace and safety belts, only one was lost: senior flight attendant Clarabelle Lansing, who disappeared into the void in the first seconds of the crisis, never to be seen again. The airframe itself was terminal. Never before or since has a civil jetliner survived such a colossal insult to its structural integrity. The incident—henceforth referred to simply as "Aloha"—is still widely remembered by aviation engineers, who, with the dark humor of every profession that grapples with tragedy, sometimes refer to it as the "737-cabriolet."

NESTED CAUSES

When the NTSB published its report into Aloha the following year, it identified the proximate cause to be a fateful combination of stress fractures, saltwater corrosion, and metal fatigue (NTSB 1989). Boeing built the skin of its early 737s from layered aluminum sheets, bonded together with rivets and an epoxy glue. The glue came on a "scrim" tape. Assembly workers would keep the tape refrigerated until it was in place and then cure the glue by gradually letting it warm. This was a delicate process, however, and if the tape cooled at the wrong speed, the glue would cure incorrectly. Even under the best conditions, it would occasionally bind to oxide on the surface of the aluminum rather than the metal itself (Aubury 1992). It was relatively common for the bonding to be less than perfect, therefore, and the NTSB concluded that this had been a causal factor in the accident. The board's investigation found that imperfect bonding in Aloha's fuselage had allowed salt water to creep between its aluminum sheets. Over time, this had corroded the metal, forcing the sheets apart and creating stress around the rivets. The stress in turn had fostered fatigue cracks in the fuselage, which eventually caused it to fail, suddenly and catastrophically.[1]

The causes of accidents are invariably nested, however, and the NTSB's technical explanation of Aloha raised a series of deeper questions. Structural failures of this magnitude were not supposed to happen so suddenly, whatever the underlying cause. It was not considered possible. Experts had long understood that some 737 fuselages had imperfect bonding, and this could induce stress fatigue, but they also understood that fatigue cracks progressed gradually—slowly enough that routine inspections should have raised alarms long before cracks could cause any kind of rupture.[2] Because of this, the NTSB assigned much of the culpability for the failure to the airline's "deficient" maintenance program, which, when examined closely, proved to be more unruly in practice than it should have been in principle (NTSB 1989, §2.3).

The airline contested this finding, insisting the NTSB report gave a false impression that its maintenance practices were untypical (Cushman 1989). It is easy to be skeptical of such protestations, but the argument seems credible in hindsight. As we have seen, close examinations of technological practice routinely find it to be more untidy and ambiguous in practice than in theory, and this goes as much for maintenance as for design (e.g., Langewiesche 1998b). Few would deny that an airline's maintenance practices could, in principle, be deficient to a point where they became legitimately negligent, but there is little evidence to suggest that Aloha's met this benchmark. Even the NTSB agreed that the airplane was compliant with all relevant FAA mandates, and, tellingly perhaps, the accident led to very few sanctions on the airline.

The official interpretation of the accident as a "maintenance failure" is complicated further by the fact that the 737's fuselage was supposed to be

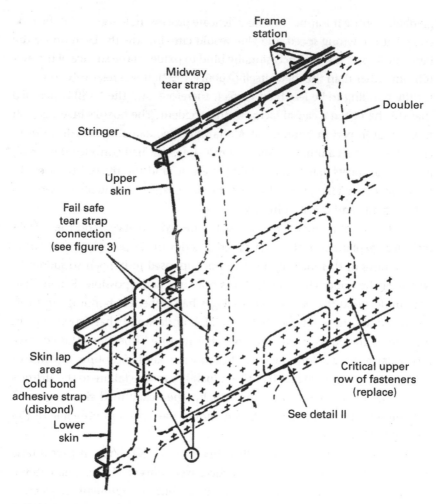

Frame station

Midway tear strap

Doubler

Stringer

Upper skin

Fail safe tear strap connection (see figure 3)

Skin lap area

Cold bond adhesive strap (disbond)

Lower skin

Critical upper row of fasteners (replace)

See detail II

①

FIGURE 7.2
Boeing 737 fuselage. *Source:* NTSB (1989).

safe even if maintenance inspections fell spectacularly short. This belief was premised on the "fail-safe" design of its metal skin, which was built—per certification requirements—to be tolerant of crack-induced ruptures and breeches. Boeing had achieved this by dividing the skin into small—10-by-20-inch (25.4-by-50.8-cm)—rectangular panels (known as "frame-bays"), each bounded by "tear straps" designed to constrain breaches by channeling and redirecting cracks, much as perforations in paper control tearing (figure 7.2). In theory, therefore, any rupture in the fuselage should have caused it to "flap" open around a single panel, releasing the internal cabin pressure in a way that was limited, controlled, and—importantly—not threatening to the airframe's basic integrity (NTSB 1989, 34). For extra security, the design allowed for cracks of up to 40 inches that encompassed two panels simultaneously.

Fully understanding Aloha, therefore, requires an explanation that reaches beyond the airframe's maintenance and accounts for the failure of its fail-safe design. And herein lies the accident's most generalizable insights. The story of how this theoretically impossible failure was possible illustrates why the finitist limitations of tests and models translate directly into limits on technological safety. In doing so, it suggests a new perspective on why catastrophic technologies fail despite our best-laid plans, and it lays an important foundation for understanding why jetliners, specifically, fail so infrequently.

Before looking further at Aloha, however, it is worth contextualizing the accident by looking briefly at the broader academic literature around technological disaster.

7.2 THE LIMITS OF CONTROL

FAILURES OF FORESIGHT

If a complex technology like a jetliner fails, especially if it fails in a manner thought to be impossible, then it is easy to interpret the failure as some kind of error: be it in design, manufacture, oversight, maintenance, or operation. A core tenet of what Jasanoff (2005) calls our "civic epistemology" of technological risk—our shared expectations about the kind of problem that it poses, and the broad manner in which it should be governed—is that all technological accidents are (in principle at least) avoidable. This understanding reflects the positivism implicit in mainstream conceptions of engineering knowledge. We believe that there is something ontologically distinctive about *failing* technologies—something that sets them apart from *functioning* technologies, which proper procedure should (or, at minimum, could) always identify in advance if experts were appropriately organized, skilled, and incentivized.

Professional engineering discourse reifies this belief by routinely portraying technological disasters as events that were *allowed* to happen, albeit unintentionally.[3] Media discourse follows the same template: framing catastrophic accidents in terms that emphasize their fundamental preventability

(Downer 2014; Hilgartner 2007). And, as discussed earlier, the academic literature on disaster usually reflects the same implicit conviction, especially in when discussing accidents born of design weaknesses. A lot of this scholarship touches on the difficulties of knowing machines. As early as 1976, for example, Turner (1976, 379) was highlighting, as a practical matter, the limited data and theory with which engineers routinely operate, as well as the implications of this for safety (see also Turner and Pidgeon 1997; Weick 1998; Vaughan 1996). But even in its most nuanced forms, the balance of this literature very rarely "bites the bullet of uncertainty" as Pinch (1991, 155) puts it, by articulating and exploring the *necessary* and *unavoidable* limits of knowledge and the significance of those limits for understanding disaster.

As we saw, however, there is one significant exception to the general assumption that technological failures are theoretically (if not always realistically) avoidable: Perrow's (1986, 1999) Normal Accident Theory (NAT). Perrow explicitly rejects the idea that perfect organizational practices (should they be possible) could yield perfectly reliable machines. And for the purposes of contextualizing the argument that follows, it is worth pausing to examine his argument in more detail.

NORMAL ACCIDENTS

NAT, developed by Yale sociologist Charles Perrow, is most fully articulated in his book *Normal Accidents* (Perrow 1999 [1984]). The text is wide-ranging and accommodates multiple interpretations (see, e.g., Le Coze 2015), but a useful way to understand the theory's core thesis is as an argument about probability and the taming of chance. By this reading, Perrow contends that some accidents in systems with certain properties—his eponymous normal accidents—are fundamentally unforeseeable and unavoidable because they stem from coincidences that are too improbable to identify in advance.

At the heart of this argument are two deceptively simple insights. The first is that accidents can result from fatal one-in-a-billion confluences (which no analysis could ever anticipate) of otherwise unremarkable anomalies (of a kind that no design could ever avoid entirely),[4] which compound each other to create a catastrophe. Perrow argues, for instance, that the 1979 accident at Three Mile Island exemplifies this phenomenon. By his account, the accident began when leaking moisture from a blocked filter tripped valves controlling the plant's cooling system. Redundant backup valves should have intervened, but they were inexplicably and erroneously locked closed. The closed valves should have been clear from an indicator light, but it was obscured by a tag hanging from a switch above. A tertiary line of technological defense, a relief valve, should then have opened, but it also malfunctioned (which went unnoticed because a different indicator light simultaneously failed, erroneously indicating that the relief valve was functioning). None of these failures or anomalies was particularly noteworthy in itself, he argues, but together they created a catastrophic situation that controllers understandably struggled to comprehend (Perrow 1999).

Perrow's second insight is that these kinds of fateful one-in-a-billion coincidences are statistically probable in systems where there are billions of opportunities for them to occur: specifically, those that are composed of many closely interconnected and highly interdependent elements. (Systems that are "interactively complex" and "tightly coupled," in his terminology.) Because where a system features many interactions between different elements, the number of unanticipated anomalies (and combinations of anomalies) that may occur in it is greatly magnified. And where its safe functioning depends on those elements all working together, the difficulty of managing such unexpected anomalies is greatly increased (Perrow 1999).

So it is, he argues, that certain types of system unavoidably harbor the potential for accidents that escape even the best oversight and control mechanisms: ghosts that lurk in the interstices of engineering risk calculations. Because where potentially dangerous systems have millions of interacting elements that allow billions of unexpected events and interactions, then seemingly impossible billion-to-one coincidences—too remote to register in any engineering analysis—are only to be expected (i.e., they are "normal").[5]

Perrow's argument is exceptional in the context of disaster research because it speaks to the inherent limits of engineering knowledge. It clearly and unambiguously articulates a case for why accidents can occur *without meaningful error*, such that their causes resist all organizational explanations and remedies.[6] NAT, we might say, delineates a conceptually important category of accidents that experts could never even aspire to organize away. It does not claim that all accidents have this character. Indeed, Perrow (1999, 70–71) argues that normal accidents are rare, and that most accidents are potentially avoidable "component failure accidents," characterized by predictable relationships between elements, or by a single, catastrophic technological fault rather than a combination of faults across a system. ("Normal," in this context, is intended to connote "expected" rather than "common.")

Perrow does not believe that Challenger, Bhopal, or Chernobyl were normal accidents, for example. He interprets them in a traditionally positivist fashion: as a product of errors, some of which were culpable, and all of which might, in theory at least, have been organized out of existence.

RATIONAL ACCIDENTS

Understanding the basic contours of NAT is useful for framing Aloha because, construed in Perrow's terms, Aloha—the failure of a single critical element (the fuselage) from a known cause (fatigue)—was a quintessential component failure accident. Even according to NAT, therefore, it should have been avoidable. If engineers were adequately rigorous with their tests, thorough with their inspections, and assiduous with their measurements, NAT suggests, the accident would not have happened.

Understanding the accident through a finitist lens, however, suggests a different view. The preceding chapters of this book have explored and illustrated the argument that expert understandings of technical systems necessarily hinge on qualitative interpretations and judgments (e.g., about the representativeness of tests and models). These interpretations and judgments can be better or worse, considered or rash, skillful or inept; what they cannot be, however, is "knowably perfect." There are infinite ways in which a test or model might be unrepresentative, and it is impossible to examine them all. If even the most rigorous engineering analyses are imperfect, however, then it cannot be true that flawed technologies are always distinguishable from flawless technologies. This is axiomatic. If experts cannot know the accuracy of their tests and models with certainty, then they cannot use those tests and models to know (and thus predict and/or control) a system's failure behavior with certainty. It is logical to conclude from this that accidents could result from conditions caused by (and unrecognized because of) erroneous, but nevertheless *rationally held*, engineering beliefs. And that, like normal accidents, these accidents would be fundamentally unavoidable.

Elsewhere, I have referred to such accidents as "epistemic accidents" (Downer 2011b; 2020). "Epistemic" can be an onerous word, however, so in this volume I will refer to them instead as *"rational accidents."* (The change in label is not intended to connote any meaningful change in definition.)

Rational accidents can be defined as accidents that occur because a technological understanding proves to be unsound, even though there were rational reasons to hold that understanding before (although not after) the event. Unlike normal accidents, which arise at the system level, from the indeterminacies of interactions between elements, rational accidents can arise from uncertainties embedded in a single element. Like normal accidents, however, they are inherently unpreventable and unpredictable. Investigations into their causes would find unproven assumptions underpinning the flawed system's design, but so too would investigations into fully functional systems, so it is wrong to imagine that the former are ontologically distinct from the latter. (To paraphrase Bloor [1976], our understandings of each should be "symmetrical.")

With this idea in mind, let us now return to Aloha, which I will argue is an exemplary rational accident. It occurred because expert understandings of the airplane's fuselage were incomplete, and this incompleteness is more appropriately attributed to the epistemological limits of proof than to any insufficiency of effort, rigor, or logic.

7.3   ALOHA REVISITED

MULTIPLE SITE DAMAGE

To understand how Aloha might be construed as a rational accident, it helps to begin by understanding more about Multiple Site Damage (MSD): the almost imperceptible fatigue-cracking that ultimately led the fuselage to fail so spectacularly.

MSD was a known problem at the time of the accident, but neither Boeing nor the FAA considered it a major safety issue. Their lack of concern in this regard stemmed, in large part, from a belief—conventional across the industry—that no MSD crack could grow from a microscopic level to 40 inches (the level to which the fail-safe tear-panels had been tested) in the period between maintenance inspections. As it was understood at the time, MSD should always have manifested as cracks that were detectable (or, at absolute minimum, controllable by the tear panels), long before they became dangerous.

As Aloha demonstrated, however, this understanding of MSD was dangerously flawed. Specifically, it failed to recognize that in certain areas of the fuselage, and under certain conditions (specifically, where there was significant disbonding between fuselage sheets, combined with a corrosive saltwater environment and a significant passage of time), MSD had a tendency to develop along a horizontal plane between a line of rivets (figure 7.3). And,

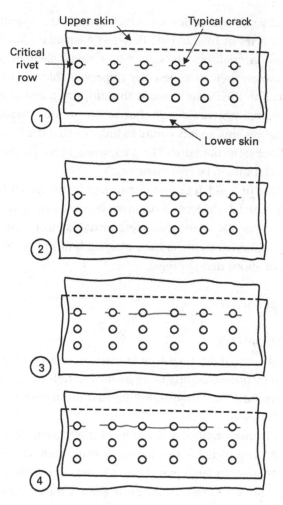

FIGURE 7.3
MSD cracks. *Source:* NTSB (1989).

further, that such a string of almost indiscernible damage could abruptly coalesce into one huge crack, longer than 40 inches, that would nullify the fail-safe tear straps and, in Aloha's case, almost bring down an aircraft (NTSB 1989: §2.3). "The Aloha accident stunned the industry," as the FAA's "Lessons learned" website puts it, "by demonstrating the [unforeseen] effects of undetected multiple site damage."

This widespread misunderstanding of MSD raises its own questions, however, and itself deserves exploration. After all, it seems reasonable to imagine that engineers should have understood how fuselages fatigue.

DARK ARTS

To understand engineers' relationship to metal fatigue at the time of Aloha, it helps to know a little history. Organized research into metal fatigue began in the mid-nineteenth century, when it first became apparent that seemingly good machine parts were failing as they aged. By 1849, engineers had coined the term "metal fatigue" and were actively working to better understand the problem; and by the 1870s, they had carefully documented the failure behavior of various alloys, even though the causes of this behavior remained opaque (Garrison 2005). Over time, the formal study of metal fatigue grew steadiliy from there, eventually coming to be known as "fracture mechanics." Its practitioners examine the properties (principally the tensile strength and elasticity)[7] of metals, as well as their relationship to different kinds of stress.

Fracture mechanics has long been a central concern of aeronautical engineering, having shot to prominence at the dawn of the jet age when fatigue felled two of the world's first jetliners—de Havilland Comets—within a four-month period of 1954 (Faith 1996, 158–165). "Although much was known about metal fatigue," Geoffrey de Havilland would lament in his autobiography, "not enough was known about it by anyone, anywhere" (quoted in Marks 2009, 20). Having borne witness to the dangers that could hide in uncertainties about fatigue, the aviation industry worked hard to explore the phenomenon. By the time Aloha occurred in 1988, therefore, its understanding of the subject was grounded in extensive experiments and decades of experience with aluminum airplanes. As the accident reaffirmed, however, this research and experience had not perfected that understanding. Even today—over sixty years after the Comet accidents, and over thirty since Aloha—aircraft fatigue management remains an inexact science, one that, in 2005, after fatigue felled a Royal Air Force transport aircraft outside Baghdad, *Air Safety Week* (2005) described as "a dark black art . . . akin to necromancy."

The intransigent uncertainties of metal fatigue are rooted in complexities that resonate with those of bird-strike tests. In the laboratory, where minimally defective materials in straightforward forms are subjected to known stresses, fracture mechanics is a complex but broadly manageable science. Decades of theory and experimentation have led to models that work tolerably well in predicting where, when, and how a metal form will fail, even if the most accurate models must grapple with quantum mechanics. In real, operational aircraft, by contrast, the problem of relevance asserts itself and the representativeness of the laboratory models becomes questionable. In this more practical realm, elaborate geometric forms, replete with irregularities and imperfections, experience variable and uncertain stresses, the vicissitudes of time, and the insults of human carelessness. All this introduces uncertainty. In these circumstances, even modest and seemingly innocuous design choices can create unanticipated stress points, as can minor deformations like scratch marks or defects in the metal itself. Such variables are so challenging to accurately model in advance that, by some accounts, even the most sophisticated fatigue predictions about real fuselages essentially amount to informed guesses (Feeler 1991, 1–2; Gordon 2018 [1991]: locs. 853–854, 1236–1239).

Compounding this complexity, the properties that make fatigue difficult to predict also make it difficult to monitor. Unlike most metals, which do not fatigue until they reach a specific stress threshold, the aluminum alloys used to make airframes (in the past at least; the industry has recently embraced advanced composites, as we will see) are thought to fatigue at any stress level. This precrack fatigue was once believed to develop at the microscopic level of the metal's crystalline structure, but now (post-Aloha) is understood to accumulate, in a nonlinear fashion, at the atomic level, where it is functionally invisible to maintenance engineers. (As of 2008, "nondestructive evaluation" techniques could detect cracks only as small as 0.04 inch [Maksel 2008]). The result, in the words of one engineer, being that "until cracking begins, it is for all practical purposes impossible to tell whether it will begin in 20 years, or tomorrow" (Garrison 2005).

DUHEM-QUINE

The complexities of fatigue might explain why engineers failed to predict Aloha's vulnerability to MSD and the inadequacies of its tear panels, but it is less obvious why those failed predictions survived the 737's extensive compliance tests. Before the airplane entered service, Boeing, under FAA supervision, extensively tested the airframe's fatigue resilience and decompression behavior; including the efficacy of the tear panels. And years later, as some 737s approached their original design life, the company even acquired and retested an old airframe. Why, then, did the dangers go undetected?

The answer, in essence, is that Boeing's tests of its fatigue predictions were framed by the same imperfect theory that framed the predictions themselves. Engineers use tests to examine the validity of their beliefs about a system's functioning, but, as we have seen, tests are themselves inescapably theory-laden (in that they embody complex ideas about the relevance of different variables, and so on). To test one theory, therefore, is to always stand on another: there can be no view from nowhere. Epistemologists refer to this dilemma—where theories cannot be tested independently from other theories—as the "Duhem-Quine Problem," and Aloha elegantly illustrates its real-world implications.

The 737's fatigue tests were premised in part on a belief that the key determinant of fatigue in an aircraft fuselage was not its age or its hours of service, but its number of "pressurization cycles" (every takeoff and landing usually constituting one full cycle). This belief was so deeply embedded in the industry's understanding of fuselage fatigue—so much so that the "fatigue life" of an airframe was expressed as a number of "cycles" to failure (NTSB 2006b, 87). It was logical, therefore, that 737's fatigue tests should be framed in the same terms. Testers simulated service experience by pressurizing and depressurizing (known as "cycling") a fuselage half-section 150,000 times (representing twice the airplane's design life). This produced no major MSD cracks and fulfilled all FAA certification requirements (NTSB 1989: §1.17.2).

In Aloha's specific case, however, the airframe's pressurization cycles was not the only factor relevant to understanding its fatigue behavior. The airplane was certainly "highly cycled" (because of its short routes), but a range of other conditions contributed to the MSD that brought it down. One was its operating environment: the warm, saltwater air around Hawaii. Another was the manufacturing flaws in its structure: the imperfect bonding in its fuselage. A third was its sheer chronological age: manufactured in 1969, it was one of the longest-serving 737s in operation. These three factors—environment, manufacture, and age—were all crucial to its failure. The disbonding created gaps that allowed Hawaii's salt water to intrude. And, over a long period of time, that water corroded the fuselage in a way that stressed its rivets and nurtured the cracks that caused it to fail. At the same time, however, all these factors set Aloha's airframe apart from the new, properly bonded fuselage that Boeing repeatedly pressurized in a dry laboratory over a highly compressed time frame. It also distinguished the airframe from the older airframe Boeing used for follow-up tests, which again had a properly bonded fuselage that had not been flying short routes in salty Hawaii. As a result, the tests failed to fully represent Aloha's real-world circumstances. By isolating pressurization as the limiting factor in fatigue, the engineers had unwittingly excluded a range of variables that would have been highly significant to predicting the airplane's failure behavior.

Note the circularity of this error, where theories about the causes of fatigue shaped the tests intended to interrogate those theories, thereby rendering those tests blind to the kinds of theoretical shortcomings that they were intended to reveal. The logic underpinning the 737's structural integrity enjoyed an internal consistency, we might say, but one that held itself aloft by its own bootstraps. This is a property of knowledge that is often highlighted by finitists (e.g., Collins 1985; Kuhn 1996 [1962]; Quine 1975). Bucciarelli (1994, 92), for instance, speaks of "The incestuous character" of what he calls the "model-making process," wherein "the model [is] designed to verify the field data; the data, in turn, providing a reference for the model."

It is difficult to fault the testers for this circularity. Boeing's (and the FAA's) engineers had run into an intractable dilemma that Collins (1985, 84) calls the "experimenter's regress." They had no way of determining the accuracy of their findings without knowing the representativeness of their tests, and no way of determining the representativeness of their tests without knowing the accuracy of their findings.

Once established, moreover, the imperfect understanding of fatigue born of this regress propagated throughout the airframe's wider design and test regimen. Importantly (and somewhat ironically), for instance, it undermined the fail-safe tear panels that were intended to serve as an ultimate hedge against errors and misunderstandings. This is to say that it led engineers to design the panels around the premise that escaped engine blades, not fatigue cracks, would cause the largest possible fuselage ruptures. And it then hid the consequences of this misconception by shaping the way that the panels were tested. Believing that escaped engine blades posed the most risk, engineers tested the panels by "guillotining" a fully pressurized fuselage section with two 15-inch blades that represented engine fragments. This created punctures smaller than 40 inches, which traversed only one "tear-strap" and led the skin to flap open exactly as predicted (NTSB 1989: §1.17.2). Unfortunately, however, it also left engineers blind to the panels' inability to cope with the kind of MSD rupture that almost felled Aloha. As the closing line of the Comet accident report had put it years earlier: "extrapolation is the fertile parent of error" (Allen 2004, 19).

7.4 IMPLICATIONS

KNOWLEDGE AND DESIGN
The story of Aloha's hidden vulnerability speaks to the unusual symmetry, in safety-critical engineering, between "knowing" and "designing." In

highlighting the catastrophic potential of an unexpected property of metal fatigue—which only became dangerous when an airframe with an uncommon manufacturing defect operated for years in a specific environment—it illustrates how even the most marginal misunderstandings can be significant to a system's failure behavior. Seen in this light, it becomes easier to understand why "technological reliability" begins to converge with "epistemological truth" as demands on the former begin to rise. And, as a result, it becomes easier to appreciate the finitist case against catastrophic technologies. The more reliability required of a complex system, the more prohibitive the problem posed by rational accidents.

It is worth noting that engineers in this domain widely recognize that they always have gaps in their knowledge, and that these gaps can be a source of danger to airplanes. "The wit of a man cannot anticipate, hence prevent, everything that could go wrong with an airplane in flight," as Newhouse (1982, 83) puts it: a sentiment that is invoked often by aviation practitioners (e.g., NAS 1980, 41) and academic observers (e.g., Mowery and Rosenberg 1981, 348) alike. Turner and Pidgeon (1997; 1978, 71ff), for instance, speak of "notional normality" in engineering, wherein, "perceptions of risk are sustained by sufficiently accurate individual and organizational beliefs about the world . . . up to the point that those beliefs are challenged by a major disaster or crisis."

Despite this recognition, however, the nature and implications of the relationship between epistemology and disaster are routinely underconsidered. It is worth exploring these implications and their significance, therefore, and a useful way of framing such an examination is to compare briefly the properties of what I have called rational accidents with those of Perrow's normal accidents.

Three properties, in particular, are worth highlighting here. I will call them "avoidability," "vulnerability," and "learning."

AVOIDABILITY As we have seen, both rational and normal accidents are fundamentally unavoidable and (in the specific rather than general sense) unforeseeable. The cause of this is different in each case, however. Normal accidents are unavoidable because engineers cannot wholly predict the multiplicity of possible interactions in a system, whereas rational accidents are unavoidable because the myriad knowledge claims that engineers draw on when designing and evaluating systems are inherently fallible.

One notable implication of this difference, as we also have seen, is that rational accidents and normal accidents define the scope of unavoidability differently, with the former suggesting that its ambit extends further than NAT would allow. Aloha was not a normal accident—there was no unforeseeable, billion-to-one confluence of otherwise foreseeable events, just the failure of a single element (the fuselage)—but Aloha nevertheless has a good claim to being unavoidable on epistemological grounds. The design of its fuselage embodied complex theories about metal fatigue; those theories were built on (and then reified by) tests and models; and those tests and models were themselves inescapably theory laden. This circularity created an irreducible measure of uncertainty: no process, however rigorous, could have guaranteed that every judgment underpinning the airframe's design and assessment was correct. Its design might have embodied flawed beliefs, therefore, but those beliefs were neither lazy nor illogical. It would be unreasonable, in these circumstances, to construe the accident as a failure of foresight.

(This is not to say, of course, that fatal engineering errors and design flaws can *never* be unreasonable or culpable. No epistemologist would deny that there are *better* and *worse* ways of establishing the properties of artifacts, even if there is no *perfect* way. Thus, there are undoubtedly "responsible" and "irresponsible" engineering practices, even if both have socially negotiated definitions and the former can never guarantee safety. The point is that *some* accidents will always be unavoidable on epistemic grounds, even if scholars might debate the extent to which any specific accident should qualify. It goes without saying, therefore, that social scientists should continue to explore the social, psychological, and organizational foundations of error. The social practices underpinning technological safety are undeniably consequential and demonstrably improvable, even if they are not perfectible.)

VULNERABILITY As with normal accidents, there are good reasons to imagine that rational accidents are more probable in some systems than in others. Because these accidents have different causal mechanisms, however, the probability of each is controlled by different variables.

As outlined previously, Perrow's key indicators of a system's vulnerability to normal accidents are "tight coupling" and "high interactive complexity." Both these properties are also likely to make systems more susceptible to rational accidents. Coupling, because the more tightly interdependent the elements in a system are, the more likely it is that any epistemologically driven failures will instigate catastrophic accidents; complexity, because every extra element in a system, and every extra relationship between elements, represents a new set of potentially fallible knowledge claims. Beyond this, however, rational accidents are likely to vary with properties that normal accident theorists might otherwise ignore.

Consider, for instance, the "variegation" of a system, which is to say the level of differentiation between its elements. Systems can be highly complex without being very variegated. Early computers, for example, consisted of thousands of identical vacuum tubes. And from an NAT perspective, such systems might be no less "complex" than those with high variegation (consisting of differently designed elements, made of dissimilar materials, performing many distinct functions). Significantly, however, highly variegated systems represent a much larger number of knowledge claims, and it follows from this that they would be significantly more vulnerable to rational accidents.

Consider also the "innovativeness" of a system, loosely defined as the extent to which it stretches the boundaries of established theory and prior experience. An airframe panel made from a new material, for instance, is neither complex nor tightly coupled, and as such would not be flagged by normal accident theorists as a source of vulnerability. (Perrow [1999, 128] actually cites the aviation industry's "use of exotic new materials" as a factor that directly contributes to the safety of modern aircraft.) From a rational accidents perspective, however, this innovation looks inherently risky. It deprives the experts charged with designing and assessing the panel of decades of research and service experience on which they might otherwise have drawn when anticipating its failure behavior, creating additional epistemic uncertainty.

(As with Perrow's complexity and coupling, the innovativeness of a system might be difficult to measure exactly or quantify objectively, as might the degree of variegation between systems, but this does not mean that these terms have no analytical value. Beauty is proverbially subjective, but this hardly negates its existence, its usefulness as an explanatory category, or its tangible consequences.)

LEARNING The aforementioned relationship between experience and knowledge speaks to a third distinction between rational and normal accidents: their different relationship to hindsight and, through it, to learning.

Take, for instance, the extent to which each kind of accident could be said to have "warning signals" that experts missed. Even with the benefit of hindsight, it makes little sense to speak about normal accidents having warning signals. The kinds of minor anomalies that combine to cause them are *expected* to occur in complex systems; it is only in their unexpected confluence that they become meaningful. So it is that normal accidents seemingly come from nowhere. (They almost have to, because if terrible coincidences built incrementally and identifiably over time, they would no longer be coincidences.) Rational accidents, by contrast, are significantly different in this respect. From the vantage of hindsight, for instance, Aloha looks replete with potential warning signals. There are key aspects of the accident—such as the planar accretion of MSD—that experts today would be expected to identify as a precursor to disaster. Such signals were not obvious at the time, because experts did not know where to look or for what they should be searching, but they are discernible in retrospect and so would constitute warning signals today. (We might even say that those warning signals *existed* only in retrospect.) The fatigue that led to Aloha would probably have been caught by modern maintenance practices, for example, but only because of insights gleaned from the accident itself. "New knowledge can turn normality into hazards overnight," as Beck (1999, 58) puts it.

It follows from this that experts might *learn* from rational accidents, by leveraging hindsight in a way that is not possible with normal accidents.

In their purest from, normal accidents yield few useful lessons. There are two broad reasons for this. The first is that the accidents from the same proximate causes are *highly unlikely to reoccur*. Where there are billions of possible billion-to-one coincidences that can instigate a catastrophic accident, then it is logical to anticipate repeated accidents, but not to anticipate the exact same accident twice. Addressing the specific factors that contributed to one normal accident, therefore, is unlikely to protect against the next. The second reason why normal accidents yield few lessons is that they *do not challenge common engineering understandings and theories about the world*. This is because the factors that combine to produce them are not surprising in themselves. For example, a stuck valve usually reveals little about valves in general and does nothing to challenge the knowledge underlying their design. The surprising aspect of normal accidents lies in the coincidence of different failures compounding each other. So it is that normal accidents only teach one lesson, and it is always the same: *experts can never design out every tragic coincidence*. This insight has important policy ramifications, to be sure, but it is of very limited value to experts looking to improve systems over time.

Rational accidents are very different in both these respects. Unlike normal accidents, the events that instigate them are likely to reoccur *if left uncorrected*. If Aloha had disappeared over an ocean, cause undiscovered, then—as with the Comets in 1954—other aircraft would likely have failed in the same way, and for the same reason. And, again unlike normal accidents, rational accidents *challenge common engineering understandings and theories about the world*. Aloha revealed fundamental misunderstandings about metal fatigue, for instance. It also revealed meaningful ways in which fuselage tests were unrepresentative of the phenomena they sought to reproduce. These two properties of rational accidents—the fact that they reoccur and the fact that they challenge conventional understandings—mean that they can yield useful design insights. Post-Aloha, for example, experts could revisit their understanding of metal fatigue and its relationship to the 737 fuselage, and this in turn meant that they could ensure that the same accident would not happen again. Engineers understood jetliners better after the accident, in other words, and jetliners are safer as a result.

So it is that a finitist understanding of failure suggests that we understand all new technologies as real-life experiments with dangerously uncertain, but ultimately instructive outcomes. This dynamic, wherein engineers might leverage hindsight for design insights, is important. Indeed, it is key to resolving the aviation paradox.

III MASTERING ULTRAHIGH RELIABILITY

Wherein it is argued:

- That experts do not manage the reliability of jetliners via formal assessment practices, as they purport. They instead leverage service experience by interrogating failures to hone a single, stable, airframe design paradigm **(chapter 8)**.
- That the idea of a stable airframe design paradigm is more credible than might be immediately apparent; and the reliability of aircraft has been dramatically lower where airframers have deviated from that paradigm **(chapter 9)**.
- That airframers are able to leverage service experience in the way they do only because of their industry's unusual structural incentives. Regulation and good intentions are insufficient explanations for civil aviation's design choices **(chapter 10)**.
- That the significance of structural incentives needs to be qualified in light of the 737-MAX crisis; and that this significance becomes most evident in relation to crash survivability, the incentives around which are different from those around reliability **(chapter 11)**.

8 PARADOX RESOLVED: TRANSCENDING THE LIMITS OF TESTS AND MODELS

On the occasion of every accident that befalls you, remember to turn to yourself and inquire what power you have for turning it to use.
—Epictetus

8.1 FLY-BY-WIRE

DIGITAL MIDDLEMAN

In 1988, a group of highly credentialed software experts challenged the UK's aviation regulator—the Civil Aviation Authority (CAA)—over its decision to certify the new Airbus A320. The group's concerns were directed at the airplane's novel design. Unlike previous jetliners, the A320's pilots would have no mechanical control over its control surfaces. Instead, they would use a joystick to interact with a computer that would interpret their input and communicate their intentions to the ailerons, elevators, and rudder.

This innovation to the basic jetliner paradigm, commonly referred to as "fly-by-wire," promised significant economic advantages. It made the airframe lighter by exchanging hydraulics for electrical elements, for example, thereby increasing the jetliner's range and efficiency. The computer-mediated flight controls that it introduced also allowed Airbus to implement flight envelope protections—software designed to prevent pilots from accidentally making dangerous maneuvers—which led to smaller (and lighter) control surfaces and facilitated a switch from three- to two-person flight crews (Twombly 2017; Beatson 1989; Pope 2008). The flight controls also helped pilots maintain optimal trim (and thus fuel efficiency) and allowed Airbus to make all its future jetliners fly in the same way, thereby reducing the costs of training pilots to operate them (Pope 2008).

The critics, however, worried that these benefits would be offset by new risks. Placing elaborate software between the pilots and the control surfaces, they argued, introduced a lot of new uncertainty into the heart of a safety-critical system. In substantiating their fears, they evoked arguments that parallel many of those outlined in the preceding chapters of this volume (e.g., on the limits of testing) to cast doubt on the new system's reliability and the validity of its assessments. As one prominent member of the group summarized:

> We don't have the technology yet to tell if the programs have been adequately tested. We don't know what "adequately tested" means. We can't predict what errors are left after testing, what their frequency is or what their impact will be. If, after testing over a long period, the program has not crashed, then it is assumed to be okay. That presupposes that they [Airbus] will have generated all of the sort of data that will come at it in real life—and it is not clear that that will be true. (Michael Hennel, quoted in Beatson 1989)

The critics concluded that the regulator's assessment criteria were logically inadequate and the safety assertions derived from those criteria were "absurd" (Norris and Wagner 1999, 45–46; see also Beatson 1989; Langewiesche 2009b; Leveson 1988).

Not surprisingly, Airbus disagreed with the critics' concerns. Even with the complex new avionics, it argued that "the safety requirement of a total breakdown occurring only once every billion hours is achievable" (Leveson 1988). The regulator itself took a slightly more nuanced line. Like its global peers, including the FAA, the CAA ultimately concurred with Airbus and certified the design. In a striking moment of candor, however, its head of avionics and electrical systems simultaneously acknowledged the critics' central argument. "It's true that we are not able to establish to a fully verifiable level that the A320 software has no errors," he told a reporter. "It's not satisfactory, but it's a fact of life" (Beatson 1989).[1]

It is unusual, to say the least, for the regulator of a catastrophic technology to publicly refer to its own safety calculations as "not satisfactory," but service experience would eventually vindicate the certification decision. The A320 family of jetliners was a commercial success and has accrued billions of service hours since its launch. Fly-by-wire avionics have become conventional in jetliners, meanwhile, and collectively have amassed many times more hours. And, although malfunctioning flight computers have, very occasionally, been a primary cause of accidents over this period—see, for instance, the 2014 crash of Indonesia AirAsia Flight 8501, discussed in chapter 6 (BBC 2015; KNKT 2015)[2]—such accidents have been too infrequent to invalidate Airbus's original reliability assurances or the regulatory decision to accept those assurances as valid. The service record of fly-by-wire jetliners demonstrates they are at least as safe as their hydraulically actuated predecessors, if not more so—just as experts predicted.

DEFYING UNCERTAINTY

The debate around fly-by-wire neatly illustrates the aviation paradox outlined in chapter 3. Expert fears about the new system might have been misplaced, but they were well founded epistemologically. (As MacKenzie [2001] and others attest, the dependability of computer systems always ultimately hinges on unverifiable human judgments [see also Butler and Finelli 1993]). Even regulators acknowledged the impossibility of fully knowing the system's behavior from their tests and models. It ought to be surprising to us, therefore, that, having been in service so long, the system has not surprised the aviation community more often. Billions of hours of catastrophe-free service suggest that its designers (and its assessors) achieved a depth of understanding that is difficult to reconcile with the irreducible uncertainties of interrogating complex machines in artificial settings. Why are there so few rational accidents, in other words, and how did experts anticipate this in advance?

This chapter will attempt to answer this question in a way that reconciles the performance of jetliners with the uncertainties of the tests and models through which they are known. The key, it will argue, lies in recognizing that experts manage jetliner reliability *actuarily* more than *predictively:* relying less on tests and models than on real-world service experience. This is to say that experts in this arena invoke the past to master the future: the historical data on jetliner reliability being constitutive of, and essential to, the reliability that it demonstrates.

The basis of this argument is reasonably straightforward, but it has underappreciated subtleties and implications. It hinges on three distinctive elements of civil aviation. The first is an organizational practice or norm that I will call "recursive practice," to which the industry shows an unusual commitment. The second is a resource that the industry enjoys in highly uncommon abundance (especially in the context of catastrophic technologies): a huge reservoir of service experience, collected under what I will call "permissive" conditions. The third is another organizational practice or norm—this one widely underappreciated in most lay discourse around the industry—which I will call "design stability." The following sections of this chapter will outline each of these elements in turn, discussing their practical value to engineers pursuing extreme reliability, their relationships to each other, and the reason why they allow experts to transcend the epistemological limits of tests and models.

Let us begin, then, with recursive practice.

8.2 RECURSIVE PRACTICE

RELIABILITY FOLLOWS FAILURE

Recursive practice—a term that I have borrowed from Constant (1999, 337)—refers to the engineering practice of leveraging insights from service experience to incrementally refine designs and understandings. All technological spheres exhibit some degree of recursive practice, and the idea of engineers incrementally whittling their uncertainty via examined experience is correspondingly well established in the innovation and engineering studies literature. Fleck (1994), for example, speaks of "learning by trying"; Arrow (1962) of "learning by doing"; Mowery and Rosenberg (1981) of "learning by using"; and Turner and Pidgeon (1997, 130) of "the restructuring of understanding." Vincenti (1990), in turn, describes complex technologies as moving from "infancy to maturity." Bohn (2005) argues that engineering and manufacturing processes gradually transform from "art" into "science." And Petroski (1992b, 2008) observes that "form follows failure" (see also Weick and Sutcliffe 2001; Wildavsky 1988, 93). But while the practical value of recursive practice is commonsensical to engineers and well documented in the literature around engineering, its absolute indispensability to achieving ultrahigh levels of reliability—and thus its criticality to managing catastrophic technologies—is less visible than it should be.

To understand the value of recursive practice to catastrophic technologies, it helps to consider its relationship to the finitist challenges of ultrahigh reliability. Chapter 7 argued that inherent uncertainties of tests and models inevitably create opportunities for unforeseeable failures: what it called "rational accidents." But it also argued that such failures can be instructive because they reveal properties and behaviors that tests and models have failed to capture. It follows from this that, over time, engineers might exploit these insights to hone their knowledge of a system's functioning, and thereby raise the reliability of their designs to levels beyond those that could be achieved with tests and models alone.[3]

Insofar as engineers can tolerate failures, therefore, recursive practice offers means of gradually transcending the finitist limitations of tests and models. Achieving extreme reliability means understanding properties and behaviors that manifest only under highly uncommon circumstances (for instance, recall from chapter 7 that the Aloha incident had a unique combination of manufacture, environment, and operation). Such properties and behaviors easily elude tests and models, but over billions of hours of real-world service they reveal themselves in failures, and engineers can use the lessons of these failures to hone and enhance their designs. So it is, as Wilbur Wright once put it, that "if you really wish to learn, you must mount a machine and become acquainted with its tricks by actual trial" (Hallion 2004, 185).

This is all to say that the first step to resolving the epistemological paradox of civil aviation's extreme reliability lies in recognizing the industry's long-standing and uncommonly far-reaching commitment to recursive practice. Observers often remark on this commitment (e.g., Mowery and Rosenberg 1981); indeed, the very notion of the "learning curve" is often traced to early discussions of aviation engineering (Wright 1936). To better appreciate it, however, it helps to think of recursive practice as consisting of two basic components: "learning lessons" and "implementing changes."

Let us consider each of these in turn.

LEARNING LESSONS   Experience cannot become insight unless it is examined, and one indicator of civil aviation's commitment to recursive practice lies in its dedication to analyzing its failures. Civil aviation is far from the only technological sphere that interrogates its misadventures. Almost every high-profile technological accident now prompts some kind of formal investigation, sometimes at considerable expense. After the 2003 *Columbia* disaster, for example, search teams collected debris from an area stretching from Texas to Louisiana. (It eventually located over 84,000 pieces of the lost shuttle, together with several meth labs and murder victims. [Langewiesche 2003]).

Even if civil aviation is not unique in working to learn from its failures, however, its sustained commitment to doing so is noteworthy and exceptional. From the earliest days of air travel, experts have endeavored to carefully investigate and interrogate aviation failures for usable insights into how to make aircraft safer. The first independent air accident investigation and report date from 1912; investigators identified pilot error as the primary cause and recommended design changes to seat restraints to mitigate future occurrences (Macrae 2014, 26). In the years since then, such efforts have become highly organized and institutionalized, garnering extensive state support. Today, major aviation incidents are investigated by specialist, publicly funded agencies with a high degree of independence, such as the National Transportation Safety Board (NTSB) in the US, which are prepared to go to great lengths to establish causes. After Air France Flight 447 crashed into the mid-Atlantic in 2009, for example, experts spent several years searching for the fuselage in an effort that cost an estimated 50 million dollars (Mohney 2014). The so-far (as of now) unsuccessful efforts to locate Malaysia Airlines Flight 370, which simply disappeared, bewilderingly, in March 2014, have cost considerably more.

Such high-profile investigations are only the most prominent facet of a much larger enterprise. Less visible, but arguably as important, are efforts by a host of organizations to collect and explore civil aviation's near-misses, lapses, and noncatastrophic failures. Macrae (2014) offers a window into this underreported work, much of it undertaken by the operators themselves, with many airlines maintaining teams of dedicated experts who collect and examine a host of safety-related incidents that arise during routine operations. These incidents are defined broadly—a typical report, for instance, might highlight flights with confusingly similar call signs and departure times (Macrae 2014, 3)—and some teams have hundreds of investigations open at any given time (Macrae 2014, 27). Parallel to this work, various public organizations also run independent incident reporting and evaluation programs. In the US, for instance, NASA operates an elaborate Aviation Safety Reporting System, designed to encourage air crews to report errors and anomalies without fear of sanction. Collectively, these efforts produce tens of thousands of data points each year, which experts mine for a spectrum of insights (Macrae 2014, 2007; Tamuz 2001, 1987; March, Sproull, and Tamuz 1991; Perrow 1999, 168–169).

We should note that such investigations are never wholly unambiguous. Much like tests and models, and for essentially the same reasons, they inevitably require investigators to make complex qualitative judgments, and it is relatively common for credible parties to contest their conclusions (Macrae 2014; Tamuz 1987; Edmondson 2011; Turner 1978). (The NTSB's explanation of Aloha has been challenged since its publication, for example, and a rival explanation proposed [*Honolulu Advertiser* 2001]). Such difficulties rarely keep investigations from offering valuable insights, however, not least because even erroneous explanations for accidents still tend to identify plausible causes, which might easily be responsible for the next accident and thus be equally useful from a reliability perspective.

IMPLEMENTING CHANGES   If one indicator of the civil aviation's commitment to recursive practice lies in the way that it seeks to learn the lessons of past failures, a second lies in the extent to which it utilizes those lessons when designing new airframes. Relative to other technological spheres (including many catastrophic technological spheres), it has unusually well-established institutional structures for mobilizing the lessons that it gleans: systematizing, promulgating, and operationalizing them across the industry. Regulators play an important role in this regard. The FAA, for instance, maintains an extensive public archive of "lessons learned" from accidents, and it has the legal authority to require design changes stemming from those lessons (even to aircraft in service already) (NAS 1998). Such powers are bolstered further by an extensive network of international agreements—many under the aegis of the International Civil Aviation Organisation (ICAO), a dedicated agency of the United Nations—that allows international bodies to share insights and (to varying degrees) enforce rules across national jurisdictions (Macrae 2014, 26, 44). For the most part, however, such lessons require minimal enforcement. The airframers themselves have established their own sophisticated practices to feed insights from service into their internal decision-making. Boeing, for example, reportedly compiles thousands of small lessons into a volume called *Design Objectives and Criteria*, which then acts as a canonical text in the planning stages of its new jetliners.

The nature of the industry's efforts to learn from past failures—how they shape airframes—will become clearer in the discussion of "design stability" that will follow, but for now, it suffices simply to recognize their scope. "The true sin" in building jetliners, write Cobb and Primo (2003, 77), is "not losing an aircraft, but losing a second aircraft due to the same cause." They are not wrong. The industry is fastidious about avoiding past errors by acting on the findings of its investigations: a huge number of the choices and judgments implicit in a new airframe design being informed by an admonitory incident. Take, for example, the two 1954 Comet accidents, mentioned earlier in this book. After painstaking (and at the time unprecedented) efforts to recover debris from the sea and re-create the fuselages on land, investigators concluded that the rectangular shape of the airframe's windows had been unexpectedly inducing fatigue cracks at the window corners. It was an informative lesson, which is manifest today in the oval windows seen on every jetliner (Faith 1996, 158–165). Langewiesche (1998b) highlights some other examples:

> An American Eagle ATR turboprop dives into a frozen field in Roselawn, Indiana, because its de-icing boots did not protect its wings from freezing rain—and as a result new boots are designed, and the entire testing process undergoes review. A USAir Boeing 737 crashes near Pittsburgh because of a rare hard-over rudder movement—and as a result a redesigned rudder-control mechanism will be installed on the whole fleet. A TWA Boeing 747 blows apart off New York because, whatever the source of ignition, its nearly empty center tank contained an explosive mixture of fuel and air—and as a result explosive mixtures may in the future be avoided.

Such stories are so common that the history of jetliner evolution can almost be written as a series of alarming events and the revelations born of investigating them. "Many a time I have had a question answered by a reference to a particular crash or incident," as one experienced aeronautical engineer put it. "We don't do this because of 'X.' We do this so we do not get a reoccurrence of 'Y.'"[4]

The FAA's regulatory requirements have followed a similar trajectory. Its certification rules and procedures have evolved in parallel to jetliner designs themselves, incrementally being updated to reflect the same hard-earned insights. So it is, for example, that the Aloha investigation shaped modern type-certification standards, as well as modern airframe designs (NTSB 2006b, 37–38).[5] The National Academy of Sciences (NAS 1980, 41) succinctly identified this process and its necessity in a 1980 review of FAA assessment practices, saying, "It is in the nature of every complex technological system that all possible risks—even mechanical ones—cannot be anticipated and prevented by the design. Most safety standards have evolved from the experience of previous errors and accidents. Airplanes built in accordance with current standards are therefore designed essentially to avoid the kinds of problems that have occurred in the past." The NTSB (2006b, 90) echoed the same point a quarter of a century later, saying, "The use of engineering analysis and tests has a long regulatory history that has produced design criteria developed over decades of flight experience. Design criteria in regulations evolve, changing as the need arises and as experience is gained with specific types of materials, components, and design features."

It is conventional to think of ultrareliable jetliner designs as being born of insightful rules and regulations, but this is misleading. The rules and the designs are better thought of as children of the same parent: rigorously examined service experience. In this context at least, Wildavsky's (1988, 2) maxim that "safety must be discovered and cannot be merely chosen" is literally true.

Learning from failures requires failures from which to learn, however, and this leads to another condition of civil aviation's success: its service experience.

8.3 PERMISSIVE SERVICE

BORN OF ADVERSITY
In 2005, a Boeing 777 out of Perth en route to Kuala Lumpur experienced a dangerous "uncommanded extreme pitch excursion" (i.e., its nose pitched up violently and suddenly, with no input from the pilot, causing the airplane to lose its lift). Investigators would later determine that the alarming anomaly was caused by a configuration error in the flight computer, which had been present since the first 777 entered service a whole decade earlier (ATSB 2007).

The airplane landed safely in this instance, but the long-undetected error that might have brought it down illustrates some constraints faced by engineers who would use recursive practice to achieve extreme reliabilities. The insights that they need to achieve this relate to failure modes that are so specific and improbable that they elude tests and models, but decades can pass before these modes manifest in actual service. So honing a system's reliability to ultrahigh levels of reliability in this way requires a huge amount of service experience from which to learn. And, since the system in question will be failing during this period, it also requires that the public be willing to tolerate it being less reliable than is ultimately desired. For want of a better term, let us call this experience collected under forgiving conditions "permissive service." Achieving the levels of reliability required of catastrophic technologies, we might say, demands an extraordinarily deep well of permissive service.

As we have seen, civil aviation is enormously privileged in this regard, almost uniquely so among catastrophic technologies. With so many aircraft flying so many routes for so many years, the industry has amassed service experience to a degree that few, if any, other technologies of comparable complexity and criticality can match. And, as we have also seen, it has done so under historical conditions that, compared to today, were extremely permissive with regard to accidents. (Partly because of changing attitudes to risk and air travel, and partly because societal concern tends to respond more directly to the absolute rather than the relative frequency of accidents—such that a higher number of accidents per departure was more tolerable when flights were less frequent.)[6]

The extraordinary failure permissiveness of civil aviation's early years, and the value of the lessons they yielded, are both evident in the industry's rate of accidents per million departures. This statistic has been declining since the dawn of powered flight: precipitously at first, and then more gradually as the industry honed its designs and understandings (see, e.g., Airbus 2017). Chapter 3 outlined the remarkable safety record of modern air travel, but this modern safety is wholly at odds with the early days of aviation. At that time, aircraft were not a dependable means of getting around. Of the nine original members of the Wright exhibition team, five were killed in accidents (Kirk 1995, 240). By 1913, the budding general aviation community was averaging about one death for every 3,000 flights (Holanda 2009, 25). A synopsis of UK civil flights over the six months from May to November 1919, meanwhile, showed thirteen accidents for every 4,000 hours flown: a rate of one for every 307 flight-hours (Chaplin 2011, 76). By 1929, the accident rate for commercial aviation was down to about 1 in every million miles traveled (Khurana 2009, 150), a figure that, if applied to 2018 data, would imply roughly 43,000 accidents per year. The risk of traveling by plane in the US around the same time has been estimated as being 1,500 times greater than traveling by rail and 900 times greater than by bus (*Air Safety Week* 2001). Commercial air travel retained an aura of being sporting for many years: passengers routinely applauding safe landings as if they were an achievement of note. It is only in

the later decades of the twentieth century that it started accruing safe landings with the remarkable frequency that we take for granted today.

So it is, as the director of the FAA testified to Congress in 1980, that
"[u]nderstanding how aircraft work was not a year's process, but a fifty- or
sixty- or seventy-year process" (US Congress 1996, 63). And accidents were
an unavoidable element of this process. Safety scholars sometimes specu-
late about averting disaster by learning intensively from near-misses and
close calls. But while such efforts are undoubtedly worthwhile, any hopes
of forestalling all accidents in this way are unrealistic. As we have seen, it is
the nature of rational accidents that they do not always exhibit identifiable
warning signs. The painful early decades of unreliable air travel should not
be understood as an unfortunate epiphenomenon of immature design or
assessment practices, therefore, but as a *necessary precondition* of the industry's later successes: an indispensable resource, without which its current
levels of safety would not have been achievable. "One cannot create experience," as Albert Camus once wrote, "one must undergo it."

Together with its commitment to recursive practice, therefore, civil aviation's deep well of permissive service goes a long way toward resolving the
paradox of its reliability. Still, however, these two conditions are not sufficient. This is because they do not account for the fact that the industry builds
new types of jetliners, and those designs have to be known to be reliable *before*
they enter service. It might have been tolerable for the industry's reliability
to improve over time, but it would not be acceptable for each new jetliner to
start out much less safe than its predecessors. To fully understand how experts
have navigated the aviation paradox, therefore, we must recognize one final
condition: a distinctive commitment to what I will call "design stability."

8.4   DESIGN STABILITY

INNOVATIVE RESTRAINT

Straightforwardly put, "design stability" refers to a practice whereby engineers evolve the designs of a complex system evolve very slowly and cautiously, with each "new" design being a modest iteration of designs already
in service. The significance of this practice to unraveling the aviation paradox stems from the fact that the kinds of errors that elude tests and models
cause rational accidents are often highly design specific, frequently relating

to very particular configurations of forces, environments, materials, and forms. Recall again, for instance, that Aloha's failure arose from a relationship between the airplane's distinct manufacturing processes, fuselage design, and operating environment.

This specificity is important because it drastically limits the "generalizability" of the lessons that engineers use to hone a system's reliability beyond levels that they could achieve with tests and models alone. It means that the hard-won insights that experts glean from accidents are likely to become irrelevant if the design changes. (For instance, most of Aloha's lessons about fatigue are being devalued as manufacturers abandon aluminum fuselages.)

Where extreme reliability is the goal, therefore, a system cannot change too dramatically or too quickly if engineers are to effectively leverage service history with recursive practice. "In order to climb the learning curve effectively and profitably, changes must be small," a former engineer for a major airframer put it, adding that "the new widget needs to look a lot like the old widget. It needs to be installed similarly, troubleshot similarly, and repaired similarly. Wholesale changes upset the system and require a total state change. State changes are expensive. State changes often cost lives."[7] Hence the commitment to design stability. We might think of modern airframes as the result of a specific form of "path-dependency" (Unruh 2000; David 1985; Arthur 1989, 2009), wherein early design choices become locked in, simply because experts accrued enough experience with them to encounter, and thus engineer out, many of their unanticipated surprises.

As with civil aviation's commitment to recursive practice, the distinctiveness of its commitment to design stability is more a matter of degree than of kind. All complex technologies evince what Kaldor (1981) calls "design inheritance," and many experience periods of relative stability (Abernathy and Utterback 1978; Kemp, Schot, and Hoogma 1998, 182). Few, if any, however, exhibit these qualities to the same extent as civil aviation, or even close. Engineering historians sometimes distinguish innovations that challenge fundamental design paradigms from those that represent more modest alterations. Vincenti (1990), for instance, contrasts "incremental design," where engineers draw on familiar concepts, with "radical design," where engineers must develop new knowledge (see also Petroski 1994). Expressed in Vincenti's terms, we might say that civil aviation emphasizes "incremental design" to a degree that sets it apart from almost any other technological sphere.

The degree of this stability of jetliner designs is admittedly counterintuitive. Manufacturers routinely market their new aircraft as "revolutionary," and it is, of course, the point of new jetliners that they differ in meaningful ways from their predecessors. Examined closely, however, the underlying architecture of jetliners has been extraordinarily stable for more than a half-century. To a degree that can be genuinely surprising, new airframes are almost always modest evolutions of their predecessors; each representing an incremental step on what Mowery and Rosenberg (1981, 349) call a single "technological trajectory." Many elements have remained unchanged for generations of aircraft (belt buckles are a very visible example, but there are many more). With one very notable exception—Concorde, which we will visit in detail in the next chapter—it would be fair to say that today's jetliners are, in many significant respects, yesterday's jetliners. It is not for nothing that journalists routinely chide the industry for not being exciting in its new offerings (e.g., Gapper 2014).

As well as being counterintuitive, this stability is difficult to substantiate, not least because "innovativeness" will always be a somewhat ambiguous, interpretive, and relative property of artifacts. For this reason, chapter 9 will explore and substantiate it in more depth. For now, however, let us simply note that historians of civil aeronautics routinely highlight the endurance and universality of the modern jetliner paradigm. By most accounts, the earliest days of civil aviation were marked by a few major design upheavals, which, about midcentury, gave way to the continual refinement of a common design that we see today. Loftin (1985) illustrates this well. He describes several waves of innovation through the early 1930s,[8] and another at the beginning of the jet age in the 1950s, followed by a long and sustained period of incremental adjustments and alterations (see also Rae 1968; Miller and Sawers 1970; Abernathy and Utterback 1978; Constant 1980; Vincenti 1990, 1994, 1997; Mowery and Rosenberg 1981; Wanhill 2002).

Even a cursory glance at an illustrated aviation timeline serves to underline this point. The first powered airplane, the Wright Flyer, is immediately and unmistakably distinguishable from the Boeing's first jetliner, the B707, which took flight fifty-five years later. The B707, by contrast, looks recognizably modern. Absent the engines—which followed a separate design trajectory, undergoing an independent revolution in the early 1970s (Smith and Mindell 2000; Constant 1980)[9]—a casual observer might struggle to distinguish

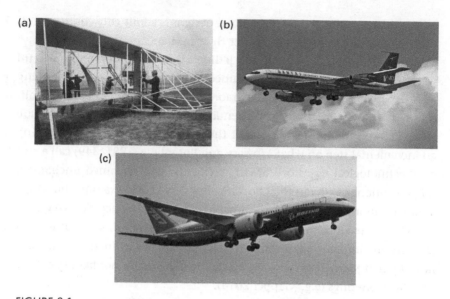

FIGURE 8.1

(a) The Wright Flyer (1903); (b) Boeing 707 (1958); (c) Boeing 787 (2011). *Source:* Wiki Commons.

the B707 from the B787, which entered service in 2011, roughly as many years after the B707 as separates the B707 from the Wright Flyer (figure 8.1).

Jetliner silhouettes are superficial evidence to be sure, and airframes have unquestionably evolved in many ways over this period, but the pattern is more than skin deep. The B707's outward similarity to its descendants (and also, significantly, to its competitors and their descendants) is indicative of something meaningful. In most cases, the internal elements of a modern jetliner—such as structural joints, fastenings, and window shapes—would be difficult to distinguish from those of a different jetliner, and often from those of much older aircraft. New variants of existing airframe types—usually denoted by a second number, such as the Boeing 737–100; 737–200, 737–300, and so on—are even more restrained. They vary only in strictly circumscribed ways from their predecessors, even to the point of eschewing innovations that have become entirely commonplace and conventional on newer airframes.

It is important to understand that design stability is not only an expression of the extent of innovation, but also of the rate at which, and the manner in which, it occurs. Civil airframers, we might say, are the old-school Tories of the engineering world: they believe in progress, but only at a modest pace that consecrates traditions and builds on the hard-earned wisdom of their predecessors. While they do innovate, therefore, they do so extremely cautiously and circumspectly. So it is, for example, that new avionics—which have arguably undergone the most frequent and dramatic changes over the last few decades—are often installed with redundant backups of a more traditional design. New materials, meanwhile, are introduced gradually across generations of airframes and implemented first in areas that are unlikely to be safety-critical (the following chapter will illustrate this point in more detail).

It is extremely rare, moreover, for airframers to use any new material, component or design concept before it has first established an extensive, and closely examined, service record in another domain. Military aviation, for instance, has been extremely important in this regard. Western air forces have provided what Kemp et al. (1998) refer to as a "strategic niche," where technologies that are new to aviation have been able to develop and mature. The military's much higher tolerance for risk allows it to employ relatively untried materials and concepts at scales that start to generate meaningful service data (Rogers 1996; Mowery and Rosenberg 1981, 355; Kemp et al. 1998, 183).

This commitment incrementalism is invariably present even in the most radical-appearing design shifts. Take, for example, the A320, outlined previously. Despite the misgivings that greeted its fly-by-wire avionics, they were hardly novel when the airplane first took flight in 1987. Military aircraft had been employing electrically driven flight controls in varying forms of sophistication since the 1930s, when Soviet engineers incorporated them into the Tupolev ANT-20 (Guston 1995); In the 1950s, Avro Canada integrated a more fully realized fly-by-wire system into a military interceptor: the CF-105 Arrow. A similar system was integral to a British experimental jet, the Avro 707B, designed in the early 1960s, and was tested extensively by NASA around the same period. Western militaries had also been employing electronic flight protections since the late 1950s; their development was motivated by collision avoidance problems arising from high-speed, low-level flight (Cowen 1972, 193).

From the mid-1970s, almost all new US fighter aircraft were fly-by-wire, in the sense that they employed both electrically driven flight controls and flight protections (Rhodes 1990). The F-16 Falcon (which first flew in 1974), and F/A-18 Hornet (from 1978)—backbones of the US air force and navy, respectively—were both fly-by-wire aircraft. The Panavia Tornado (from 1974), the cornerstone of several Western European air forces, relied heavily on active digital flight protections to stabilize and augment its flight. The F-117A Night Hawk is an extreme example. Introduced in 1981, it was inherently unstable without active computer intervention. Pilots nicknamed it "the wobblin' goblin."

By the time that the A320 launched, core elements of these fly-by-wire systems had even been edging into mainstream civil aviation. Concorde, which entered service in 1976, twelve years before the A320, had electronic flight controls. It also had the first commercial digital-electronic engine control system (Favre 1996). (Full-Authority Digital Engine Control, or "FADEC," was pioneered slightly earlier in the decade on the F-111's Pratt & Whitney PW2000 engines.)

In these aircraft and others, therefore, essential elements of the A320's "radical" fly-by-wire system had accrued many thousands of flight hours by the late 1980s. (Not to mention the countless service hours that its more generic elements—chips and servos, among others—had accrued in spheres outside aviation.) Insofar as the A320 can claim to be an aeronautical first at all, it is as the "first civilian production aircraft with digital fly-by-wire controls," where "civilian," "production," and "digital" are all necessary modifiers. And even here, there are significant caveats. The Boeing B757 and B767—both of which predate the A320, being launched in 1982 and 1981, respectively—were outfitted with a degree of digital control, including computer-activated spoilers. The A320, meanwhile, retained mechanical connections to the rudder and horizontal stabilizer (Beatson 1989). Civil airframers, an adage goes, believe that nothing should be done for the first time; and while the A320 might test this rule, it does not disprove it.

Still, however, the claim that civil airframers refrain from dramatic innovation is controversial among many industry practitioners and observers, not least because Concorde raises questions. There are good answers to their concerns, but those answers require some detailed explanations. For this reason, the next chapter will explore them in more depth. Let us park those concerns for now, therefore, and take stock.

8.5 PARADOX RESOLVED

THE RATCHET OF ULTRAHIGH RELIABILITY

Civil aviation's commitment to design stability and recursive practice, together with its extensive legacy of permissive service, offer a satisfying resolution to the aviation paradox. They do not fully explain the safety of modern air travel by themselves. As we saw, that is a much more holistic achievement with many other necessary conditions. But they do offer a way of reconciling civil aviation's extraordinary reliability with the limits implied by finitist epistemology: an explanation for how experts managed to design, and predictively assess, complex machines with a level of understanding that transcends the limited fidelity of their tests and models. They do this by offering a means of incrementally ratcheting the reliability of new jetliners (along with expert confidence in that reliability) that is grounded in real-world service experience rather than representations and abstractions.

Figure 8.2 sketches this reliability ratchet as a crude schematic, where each arrow indicates a dependency relationship. It illustrates several key processes worth reiterating and draws attention to a few important nuances of the process that might not be immediately obvious from the explanations given here.

Four points, in particular, are worth highlighting:

1. *Design stability and extensive service experience are sufficient preconditions for accurately predicting ultrahigh levels of reliability, but achieving such reliabilities also requires recursive practice.* To accurately predict ultrahigh reliability in a new system, experts require two things: (1) compelling evidence that the new system is substantially similar to a previous system (i.e., design stability); and (2) statistically significant data on how often that previous system has failed in the past (i.e., service experience). Recursive practice becomes necessary only insofar as experts are looking to improve the performance of such systems.

2. *Design stability contributes directly to the volume of useful service experience.* The greater the similarity between different designs, the more productive it is to pool their collective service data. To the extent that two jetliner types are substantially similar, in other words, then the performance of one is more likely to have statistical (and epistemological) relevance to the other.

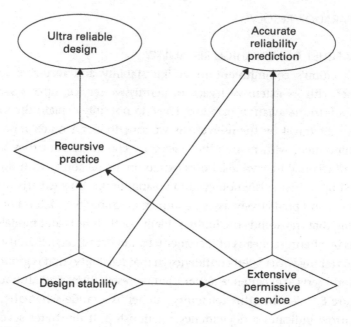

FIGURE 8.2
The ratchet of ultrahigh reliability.

3. ***Design stability contributes to the efficacy of recursive practice.*** As outlined in this discussion, the kinds of insights that escape tests and models
   tend to pertain to subtle misunderstandings of highly design-specific interactions. For this reason, they tend to lose their relevance when designs
   change. The greater the similarity among designs, therefore, the more useful recursive practice becomes for honing reliability to extreme levels.

4. ***Extensive service experience, inevitably implying a significant number
   of accidents, is a necessary precondition of achieving ultrahigh levels of
   reliability and cannot be avoided.*** Winnowing out subtle epistemological
   errors requires a deep well of experience from which to learn. This learning process is necessarily slow and painful; it implies a significant number
   of catastrophic failures from which to glean insights. Thus, it has to be
   supported by a permissive social environment, wherein such failures are
   tolerated at a higher rate than would conventionally be required of a catastrophic technology (especially what I have called an "acute" catastrophic
   technology). Experience, as Oscar Wilde once quipped, is the one thing
   you can't get for nothing.

MISREPRESENTATIONS

Note that at the heart of this resolution to the aviation paradox is the idea that experts design for, and assess jetliner reliability via techniques that differ from how they are portrayed in subtle but important ways. The paradox, as originally formulated, arose from the observation that—ostensibly at least—experts successfully use tests and models to design and assess new jetliner types, and this achievement is incommensurable with the inherent ambiguities of tests and models. Examined closely, however, the industry's experts are not building "new" aircraft so much as they are incrementally modifying one longstanding design paradigm. And the knowledge on which they are drawing to navigate that process is less grounded in tests and models than in decades of experience with operating that design paradigm, at scale, in the real world.

The reliability of modern jetliners might begin with assiduous testing and modeling, in other words, but as it has reached ever higher levels, it has increasingly been forged in the field. The industry's extensive service history revealed behaviors and conditions, some inevitably catastrophic, that its tests and models missed. Investigating those occurrences revealed insights that allowed design refinements. And keeping designs stable ensured that those refinements continued to be relevant, thereby allowing their benefits to accrue and compound over time.

The accuracy of the FAA's reliability assessments can be understood in much the same way. Regulators have been able to predict the performance of new airframes, despite the indeterminacies of their measurement practices, because there is a meaningful sense in which they are not predicting the performance of *new* airframes and are not relying solely on their measurements. Instead of assessing each system's reliability from scratch, in the laboratory and on the computer, as the certification process implies, they examine how very similar systems have performed in the past, satisfy themselves that any proposed changes to that system will not negatively affect that performance, and assume that the new system will be about as reliable as its predecessors. Contrary to appearances, therefore, their knowledge is more *actuarial* than it is *deductive*: they project the service history of old designs onto new designs, which they can do effectively because the industry does not deviate much from the common airframe paradigm it has been refining for almost seventy years, and because there is statistically significant evidence of how often it fails. The FAA's predictions are borne out *by* experience, we might say, because they are born out *of* experience. Or, as one engineer working in the industry succinctly put it: "There are no techniques available that can give you confidence in [an ultra-low] failure rate . . . unless you already know that your kit achieves that failure rate." [10]

This approach to managing and assessing ultrahigh reliability is not without uncertainties. Extrapolating from historical failure rates still requires subjective ceteris paribus judgments, for instance, and pursuing design stability requires fundamentally unprovable judgments about similarity and relevance. In practical terms, however, such uncertainties are much more manageable than those involved in inferring extreme levels of reliability directly from tests and models. The question "How can we predict the failure behavior of a substantially new system over billions of hours of real-world operation?" is much less tractable than the question "Will this specific modification—which has been closely scrutinized and analyzed and is being implemented incrementally and cautiously after extensive trials in other real-world contexts—make this system less reliable than its predecessors?" The latter is a different class of problem to the former; and one that is much easier to reconcile with finitist assertions about the limits of proof. (We might say that it returns engineering's relationship to epistemology to the pragmatic state described by Constant [1999] and outlined in chapter 2, wherein finitist arguments still hold in principle but rarely matter in practical terms.)

Understanding civil aviation's reliability practices in this way might not seem radical to many working engineers, but it runs contrary to common, widely institutionalized construals of catastrophic technologies and the work of managing them: construals that shape popular discourses, organizational structures, legal standards, and public policies. Later chapters of this book will explore these implications in more detail. Before then, however, let us pause to expand on the argument about design stability given here, which, as previously mentioned, has controversial elements that deserve unpacking.

9 DESIGN STABILITY REVISITED: CONTEXT, CAVEATS, COMPOSITES, AND CONCORDE

> You want me to tell you if this old airplane is safe to fly?
> How in the world do you think it got to be this old?
>
> —Anonymous

9.1 A CONTROVERSIAL NOTION

UNDENIABLY RADICAL

In the pantheon of jetliner designs, Concorde[1]—which first flew in 1969 and entered regular service in January 1976—is undeniably exceptional. Even standing in a museum, the airframe manages to look futuristic; its sleek lines strangely incongruous with its dated analog cockpit. This first impression is more than substantiated by the airplane's performance statistics. Concorde's cruising altitude was over 60,000 feet, which is more than 20,000 feet higher than modern jetliners. Its cruising speed was 1,354 miles per hour (Mach 2.04)—over twice the speed of sound. No modern jetliner even crosses the sound barrier. Where flights today make the trip from New York to Paris in about eight hours, Concorde routinely made it in under three and a half.

Concorde is important in the context of this argument because its existence challenges what I have called "design stability": the idea that civil aviation abhors fast, radical design changes. Chapter 8 defended the idea of design stability by arguing that the A320's fly-by-wire controls were less radical than is often remembered; the same argument cannot reasonably be made about Concorde. Even more damagingly, the airplane's service record also appears to undermine the idea that design stability is a necessary condition of modern aviation safety. It carried passengers between 1976 and 2003—twenty-seven years—with only a single catastrophic accident: Air France Flight 4590, which crashed in July 2000 while taking off in Paris, killing 113 people. (The aircraft lost a tire on the runway, fragments of which punctured a fuel tank, causing a fire that necessitated an engine shutdown. Left struggling for power at a critical moment, it rolled over and collided with a hotel [Harriss 2001]). No other jetliner has flown for so many years with so few accidents.

So it is that the history of supersonic transport poses some acute questions for my proposed resolution to the aviation paradox; and it is far from being the only reason to question the significance of design stability. Of all the claims made in the previous chapter, the notion of civil aviation being characterized by—and dependent upon—design stability is undoubtedly the most controversial. The idea is challenged further by the testimonies of people who actually design jetliners. Civil aeronautical engineers often speak about the dangers of innovation, but they also routinely object to any suggestion that their industry has not been highly innovative (or such has been the author's experience); many perceiving some mainstream jetliner designs to have been highly innovative. This poses a difficulty of its own. STS scholarship rarely sets out to challenge the technical experts that it describes, and insofar as those experts disagree with an academic representation of their work, it is usually fair to see this as a problem with the representation. They are, after all, the experts.

For these reasons, it is worth postponing the larger argument for a moment to further explore and substantiate the idea of design stability. To that end, this chapter will consider professional and academic objections to the claim that jetliners do not change radically, and to the claim that radical changes are inevitably hazardous. It will begin by examining debates around the 787's composite airframe, through which it will address the observation that many industry engineers perceive their work to be innovative. It will then look to military aviation and Concorde, through which it will address the idea that radical innovation can be safe.

Let us turn first, then, to the B787 Dreamliner and the perceptions of the engineers who designed it.

9.2   THE COMPOSITE REVOLUTION

DARK MATERIALS

In discussions about radical innovations to the conventional jetliner paradigm, two specific shifts tend to dominate. One is the introduction of fly-by-wire avionics with the A320, discussed in chapter 8. The second is the introduction of composite materials with the Boeing 787.

Various scholars, engineers, and industry observers have argued that the B787 demonstrates that major airframers are willing and able to innovate boldly. On the airplane's launch in 2011, it was widely heralded as the first jetliner to be constructed from composite materials rather than aluminum. Composites—sometimes referred to as "advanced composites" or "carbon fiber"—are matrices of fibers (predominantly carbon) embedded in an epoxy resin. Boeing embraced them in the 787 because, like fly-by-wire avionics, they offer very significant weight (and consequently fuel) savings, making the airframe more efficient and thus competitive (Marsh 2009, 16). But forgoing aluminum—and, with it, the industry's many hard-won insights that pertain to metal airframes—is widely viewed as a significant innovation to the standard jetliner paradigm.

So was the 787 radically innovative? If not, why is it perceived as such by many experts? And if so, is it less reliable than its predecessors as a consequence?

Let us deal with these questions in turn.

COMPOSITES IN CONTEXT

There is, of course, no definitive answer to the question of whether the B787 represents a radical innovation: "radical" being an inherently interpretive property. That said, however, there are compelling reasons to believe that—much like the A320, and for essentially the same reasons—it is less revolutionary than is often claimed. Examined closely, in fact, it is possible to see the airframe as exemplifying the principles of design stability as much as challenging those principles.

In essence, this is because the B787's composites, like the A320's avionics, have a much longer and more elaborate history than is conventionally remembered. First developed in the 1950s—closer in time to the Wright Flyer than the B787—composites of varying kinds had been subject to significant research activity for over sixty years by 2011. This research took place in multiple countries, but in the US, it was primarily led by NASA and the Air Force. Both organizations had conducted decades of primary research and development to explore everything from moisture absorption and radiation tolerance to long-term stress behavior and lightning resilience (Spinardi 2002; Tenney et al. 2009; Rogers 1996; Vosteen and Hadcock 1994).

As with fly-by-wire, moreover, advanced composites were introduced into airframes both gradually and cautiously. Again, this began with military airframes, after a long period of study—fourteen years by some accounts—in carefully monitored programs that started with noncritical structures. So it was that the Air Force again served as what Kemp, Schot, and Hoogma (1998) call a "niche," where the new materials could begin to mature. Both it and NASA sponsored long-running "fly and try" programs, closely monitoring and evaluating the performance of composite structures and components in service on military (and later civil) aircraft.

The rare and remarkable SR-71 Blackbird reconnaissance plane, launched secretly in 1962, was a pioneer in this respect.[2] Designed to operate at extreme speeds and temperatures, many of its secondary structural elements—about 15 percent of its total fuselage, including its wing edges, tail cone, exhaust fairings, vertical stabilizers, and inlet spikes—were made of composites (Merlin 2009, 14–17). The SR-71 was a rare and dangerous bird, however. Only thirty-two were ever constructed, over a third of which were lost to accidents.

Mainstream military aviation, by contrast, adopted composites much more incrementally. Their public debut came in 1967, in remedial structural reinforcements on the F-111 Aardvark and in a retractable speed brake for the A-7 Corsair II (Jones 1999, 39–42). Their use expanded gradually from there, beginning with other noncritical secondary structures such as fairings, hatches, and (partially redundant) control surfaces before graduating to more critical primary structures such as wings and fuselages (Tenney et al. 2009, 2). Staying with US aircraft, for instance, the S-3 Viking entered service in 1974 with a composite spoiler in its wing; the F-15 Eagle in 1976, with composite empennages (tail fins); and the F/A-18 Hornet in 1983, with composite wings and tail assembly (Tenney et al. 2009, 2; Jones 1999, 43–44). The US military's embrace of composites reached its apogee in 1997, with the introduction of the B-2 Spirit "stealth bomber": the external airframe of which is built almost entirely from advanced carbon-graphite laminates.

As military airframers grew bolder in their use of composites, civil airframers—often the same companies—tentatively began implementing them in jetliners. Again, this process began at the margins, with small-scale, intensely monitored applications in noncritical secondary structures, alongside larger-scale applications in private business aircraft. Beginning in 1973, for example, eighteen Lockheed L-1011s were fitted with composite fairings, and 109 graphite/epoxy spoilers were added to a range of B737s. Four more L-1011s were fitted with composite ailerons in 1982. And, as the 1980s progressed, a handful of DC-10s gained composite upper-aft rudders and skin panels, while ten composite elevators were fitted to B727s. Each composite element's performance was carefully tracked by the NASA Langley Service Evaluation Program (Tenney et al. 2009, 6).

By the mid-1980s, the data gleaned from this program, combined with that from Air Force operations, were beginning to offer meaningful insights into the long-term behavior of composites. This experience gave civil airframers (and their regulators) enough confidence to begin applying the materials on a more routine basis. Airbus initially led the way. The A310–300 entered service in 1985 with a carbon-fiber tail-fin, making it the first production airliner with a composite primary structure. Then the A320 was launched in 1988 with an all-composite tail assembly (Norris and Wagner 1999). Boeing proceeded in much the same way, albeit more cautiously at first. Its B777, which entered service in 1995 with a carbon-fiber tail assembly, was the first of its jetliners to feature a composite primary structure (Jones 1999, 49).[3] From there, both airframers progressively increased their use of composites with each new design. Again, the performance of these structures were closely monitored, and their application evolved over time, along with the procedures that governed their assessment, manufacture, and maintenance.

This is all to say that the "revolutionary" B787—like the "revolutionary" A320—stands at the end of a long trajectory. It is far from being the first civil jetliner to employ composites in its airframe, and far from being the first airplane to feature predominately composite wings or fuselage. Its claim to originality is as the "first wide-body civilian jetliner to feature composite primary and secondary load-carrying structures," where every modifier is significant (Jones 1999, 49). When viewed in this context, therefore, it is difficult to see the airplane as radically innovative. Boeing, NASA, and the FAA all officially contend that it represents an incremental extension of a very long-term development process: the product, as a 2009 NASA study put it, "of the significant advancements made in the understanding and use of these materials in real world aircraft" (Tenney et al. 2009, 2). And all three organizations link this incrementalism to safety.

As we have seen, however, even modest innovations can carry risks where ultrahigh reliability is required. So the question of whether the 787 represents a dangerous degree of innovation is more complicated.

DANGEROUSLY INNOVATIVE?

Despite all the heritage outlined here, some observers do consider the B787 to be dangerously innovative in its use of composites. Slayton and Spinardi (2016), for instance, attest to a meaningful subset of industry insiders who construe the airframe to be rife with unacknowledged uncertainties and hazards, arguing that these insider concerns are well founded because the "scaling up" of composites was a difficult process with unexpected complications. By way of illustration, they highlight a range of issues that arose during the airframe's manufacture and testing, such as unanticipated structural weakness, born of titanium fasteners that behaved contrary to fundamental design assumptions (Slayton and Spinardi 2016, 52–55). "[E]ven after decades of experience in the carbon fiber niche," they contend, "the process of scaling up composite components required massive levels of innovation" (Slayton and Spinardi 2016, 53; see also Mecham 2009; Norris 2009; ASSD 2009; Fraher 2014; Gates 2007). It is important that we recognize such misgivings. In debates around the B787—as in those around the A320—the claim that new jetliner designs are not radically innovative undoubtedly runs contrary to the lived experience and perception of many practitioners.

The key to reconciling such perceptions, I suggest, lies in three observations that clarify and refine the notion of design stability itself. These are as follows:

i.   That the experiences and perceptions of engineers are relative and need to be situated in their professional context.

ii. That "incremental" is not synonymous with "straightforward," especially in the context of catastrophic technologies.

iii. That design stability is a norm, imperfectly followed, occasionally misjudged, and most clearly understood in hindsight.

Let us look at each of these observations in turn.

I. PERCEPTION IS RELATIVE   To understand engineers' perceptions of innovation, it is important to recognize that "innovativeness" is an inherently subjective, situated, and relative property of technical systems. Viewed from inside civil aviation, for example, the A320's flight controls and the B787's fuselage might well be considered radically innovative. But the argument about design stability needs to be understood in a wider context: as property the industry itself. The point is that, *relative to their peers in other engineering spheres*, engineers working in civil aviation have an anomalously conservative conception of what "radically innovative" means. Viewed from outside civil aviation, in other words, the fact that some insiders perceive the 787's composites and the A320's fly-by-wire controls to be radical innovations—despite the decades of intensive study and prior service that preceded both—is anomalous and remarkable. From this perspective, those perceptions speak eloquently and emphatically to a culture of innovative restraint and a commitment to design stability in civil aviation.

II. INCREMENTAL ≠ STRAIGHTFORWARD   Claims about the radical innovativeness of new airframes often point to the difficulties and uncertainties involved in those airframes' design and manufacture. By this view, an airframe can be radical by dint of the work involved in designing it, and to speak of design stability in civil aviation is tantamount to claiming that the industry lacks ambition. (So it is, for instance, that popular accounts of engineering—aeronautical and otherwise—often portray any perceived slowness of innovation as something akin to a moral failing [e.g., Gapper 2014]).

The impulse to construe the stability of airframe designs in this way is intuitive but misleading. This is because the relative stability of airframe designs should not be understood as evidence of shallow ambition or any reluctance to tackle difficult problems; it should be understood as evidence of how difficult it is to work at extreme reliabilities, where even modest-seeming, incremental innovations pose exponentially outsized costs and challenges. The uncertainties that arise from modest design changes might be tiny, but when engineers must know a system's failure behavior with extreme certainty, over billions of hours of varied operation, they need to explore even the most marginal uncertainties, and this quickly becomes a fearsomely complicated and laborious undertaking. Engineering at the extremes of reliability, we might say, is akin to walking at very high altitudes, where the oxygen is thin and even the smallest steps take an incongruous amount of effort (which outside observers struggle to appreciate).

So it is that the B787's composites or A320's fly-by-wire avionics might be understood as *extremely challenging and ambitious innovations relative to the conventions of civil aviation*, and, simultaneously, as *extremely incremental innovations relative to the conventions of other engineering spheres*. With the gap between these understandings being evidence of civil aviation's commitment to design stability.

III. DESIGN STABILITY IS AN IMPERFECT NORM   Critics of the A320 and the B787 argued that those airplanes' respective designs were radical because they introduced underappreciated dangers. Concerns about the 787 could have some merit in some circumstances, as we will see, but insofar as they do have merit, then they reinforce, not undermine, the importance of design stability.

As discussed, fears about A320's fly-by-wire avionics turned out to be misjudged. Its critics were well justified in doubting the tests and models used to assess the system, but they overestimated both the significance of those abstractions to the assessment process, on one hand, and the radicalness of fly-by-wire as an innovation, on the other.

Fears about the 787, by contrast, are more difficult to assess. They are starting to look unfounded as the airframe accrues ever more service hours, which would fit with the argument that its composites represent a more incremental change than is widely believed. At the same time, however, many of the most serious questions around new materials concern their fatigue behaviors, which become more of an issue late in an airframe's service life. It is certainly plausible, therefore, that the scaled-up use of composites in the 787 has meaningfully diminished the airplane's reliability. Abandoning aluminum meant relinquishing a spectrum of nuanced and hard-won insights, and there are undoubtedly hazards implicit in this loss. (Take, for example, Air Transat Flight 961, an Airbus A310 that suffered a near-miss when its composite tail failed for a reason that had eluded tests and models [NTSB 2006a, 2006b; TSB 2007; Marks 2006]).[4] Given that aluminum structures kept surprising engineers even after decades of heavy service, it is difficult to imagine that composites have yielded their final insight. (As a classic text on structural engineering observes, "a deep, intuitive appreciation of the inherent cussedness of materials and structures is one of the most valuable accomplishments an engineer can have" [Gordon 2018: loc. 816–817]).

Ultimately, only time and service experience will tell if, in this instance, the industry's ambitions have superseded its caution. Norms like design stability are not binding, and it is perfectly possible that Boeing, wrestling with competing priorities, might one day be deemed to have lapsed in this regard with the 787. (Indeed, chapter 11 will argue that the company has recently fallen short in this way with its 737-MAX.) Time, as they say, will tell.

Even if the 787 does ultimately prove to be less reliable than promised, however, its critics would be vindicated, but the point about design stability would still stand. The argument of the preceding chapter is not that the industry will *never* overreach with its innovation—as we will see with Concorde shortly, it demonstrably has overreached on at least one occasion—but that doing so has consequences, and civil aviation's impressive service record reflects a willingness to avoid doing so in the past. Let us note, moreover, that those who doubt the 787's design are, in effect, endorsing the idea that innovation is a source of dangerous uncertainty to catastrophic technologies. And the fact that their fears might be justified, even after the industry's long experience with the composites, speaks forcefully to design stability's importance to reliability.

In fact, the most compelling evidence of design stability's importance to reliability comes from instances where the practice has conspicuously been ignored. Let us now shift, therefore, to examining some unambiguously innovative aircraft.

9.3 INNOVATION UNTETHERED

MILITARY ADVENTURISM

A straightforward way to see the hazards implicit in radical innovation—and, through those hazards, the significance of design stability to ultrahigh reliability engineering—is to consider occasions when aeronautical engineers have unambiguously departed from the standard airframe design paradigm. Concorde, along with its short-lived Soviet counterpart the Tu-144, are the only unambiguously radical civil jetliners to enter service in the last half-century, and we will look at them both at the close of this section. Before then, however, it is worth looking outside of civil aviation to the military sphere, which designs aircraft with very different priorities.

Unlike their civilian counterparts, military aircraft must operate in inherently adversarial environments, where dangers arise from failing to compete effectively, as well as from issues like malfunctions. For this reason, it often makes sense for military airframers to favor more radical, innovative designs that offer greater performance, even if it comes at a cost to reliability. It is better to occasionally lose pilots to design weaknesses than to have them be shot down more easily, which would jeopardize not only the pilots, but also the nation's combat effectiveness. (So it is that the Air Force prefers to speak about "survivability" rather than "safety" in this context; the former metric accounts for airplanes' ability to withstand intentionally hostile environments as well as normal operating conditions.)

For this reason, military airframers—many of which also build, or once built, civil jetliners—have long chosen to innovate far more aggressively with military airframes than with their civilian counterparts. As with jetliners, albeit in reverse, this relationship to innovation is immediately visible in the silhouettes of military airframes. B707s from 1958 might look almost indistinguishable from B787s from 2011 (figure 8.1), but nobody could confuse a 1952 B-52 Stratofortress with its 1989 successor: the B-2 Spirit (figure 9.1).

By eschewing design stability, however, military airframers have inevitably limited their ability to employ long-term recursive practice, and it follows that this should have had consequences for the reliability of their aircraft. The evidence seems to bear this out. Comparisons between the reliabilities of military and civilian airframes will always be imperfect; not least because jetfighters (or almost any military aircraft) operate in very different conditions from jetliners. (Nobody pulls nauseating high-g maneuvers in a B737—not on purpose, at least—or attempts an arrested nighttime landing on the rocking deck of an aircraft carrier in an A320.) For all the ambiguity, however, the reliability statistics around purpose-built military aircraft do make a compelling impression. Despite the hundreds of billions of dollars spent on their design and manufacture, it is difficult to dispute that military airframes have always been much less reliable than their civilian counterparts.

This disparity is reflected in a wide range of metrics. Perhaps most prosaically, for instance, it is apparent in the downtime of military aircraft, which is vastly higher than that of even the most problematic jetliners. As of September 1997, for example, the B-2 bomber required six days of maintenance—costing upward of $3.4 million per month—for every flight day (Capaccio 1997, 1). Other advanced airframes are not markedly better, and some are

FIGURE 9.1
(above) B-52 Stratofortress (1952); (below) B-2 Spirit (1989). *Source:* Wiki Commons.

arguably worse. As of 2009, the F-22 required over 30 hours of maintenance for every hour of flight (Smith 2009), and, as of 2016, the F-35 was expected to require between 41.75 and 50.1 hours (de Briganti 2016).[5] So acute are these maintenance demands that significant proportions of the US Air Force's aircraft are not mission capable at any given time. In 2017, for instance, only 71.3 percent of its aircraft were flyable on an average day, with its most advanced aircraft being significantly more affected. (Fewer than half its F-22s were regularly available, for instance, and only 52–53 percent of its B-1B Lancer and B-2A bombers [Losey 2018]).[6]

A different, albeit similarly imperfect, perspective can be gleaned from data on loss rates. In a 1995 report, the US Government Accountability Office (GAO 1995, 3) found that the US Air Force and Navy operated and maintained about one backup aircraft for every two combat-designated fighter aircraft. It deemed this number slightly excessive, but it did note—as a guideline for making predictions about the F-22—that the Air Force had lost an average of about seventeen F-16s per year to "peacetime mishaps," over the five years prior: a significant portion of its total fleet (GAO 1995, 6–7). There are few indications that the attrition rate has improved since then. Between September 2001 and February 2010, the Air Force reportedly lost "a little over" 134 aircraft outside the Central Command combat theater, many in incidents related to mechanical issues, such as when an F-15 "broke in half" during training maneuvers (Kreisher 2010).

The costs of eschewing design stability are visible not only in experts' struggles to make military airplanes reliable, but also in their failures to predictively assess the reliability of those airplanes. "The [US Air Force] generally does not know the reliability of its fielded repairable systems," concluded one PhD thesis (Hogge 2012). Take, for instance, the canopy of the F-22, which was assessed to last for 800 flight-hours between replacements (a contractual requirement), but, in the event, had to be redesigned after it lasted only an average of 331 hours in real flight (O'Rourke 2009, 24). The F-35 has been particularly confounding in this respect. A draft report from the US DoD obtained by Reuters in 2014, for example, spoke critically about an unanticipatedly high failure rate of its software (Shalal-Esa 2014). Another, obtained by Bloomberg in 2019, found that the service life of one variant, the F35B, might be about a fourth of what had been calculated (Capaccio 2019).

Let us move on, however. Because while military aviation's many dilemmas certainly suggest a tight relationship between reliability and innovation, the differing requirements and operating conditions muddy any comparisons with civil aviation. For this reason, it is more telling to consider an example from within civil aviation itself. To close, therefore, let us examine Concorde and its Soviet cousin, the Tu-144.

CONCORDE

As outlined at the beginning of this chapter, the innovativeness of Concorde is beyond dispute. As is immediately apparent from its iconic silhouette (figure 9.2), the trans-sonic jetliner was, undeniably and unambiguously, a radical departure from the standard jetliner paradigm (and thus from the norm of design stability). Engineers often speak about the discontinuities of scale,

FIGURE 9.2
Concorde. *Source:* Wiki Commons.

where changing one variable of a system or its operation—making it larger, smaller, faster, hotter, and so on—can push against hard limits like melting points or sonic boundaries and require extensive redesign. Concorde exemplifies this phenomenon.

Building a jetliner capable of cruising at twice the speed of sound forced engineers to rethink preexisting airframes and engines in fundamental ways. Aerodynamically, for instance, it led to the striking "double delta" wing shape, tailored to minimize air resistance while still providing adequate lift during takeoff and landing. Meanwhile, the extreme heat generated at high speeds required an elaborate system for circulating fuel to cool the plane's skin, along with the extensive use of novel heat-resistant materials, including an aluminium-honeycomb-epoxy composite (very different from the carbon composites featured in the 787).

As is common in tightly coupled systems, moreover, these changes further necessitated extensive downstream modifications elsewhere in the system. The unusual wings forced the plane to adopt a steep angle at low speeds, for example, which required unorthodox landing gear and a nose that drooped mechanically so that pilots could see the runway. The heat-resistant airframe would expand in flight, necessitating complex expansion joints throughout the aircraft (and subjecting them to unprecedented fatigue pressure). The new materials required that engineers rethink the airframe's decompression behaviors (which was complicated further by the unprecedented cruising altitude, itself a function of the high speed). And this went on and on: a cascade of changes, each necessitating the next (Owen 2001; Simons 2012; Orlebar 1997; White 2016).

The result, in short, was that Concorde's design embodied a spectrum of ideas, technologies, and manufacturing techniques that were new to civil aviation. (Among them were crucial elements of the fly-by-wire avionics that would characterize the A320 decades later.) Very unusually, some of these were new to all aviation spheres, military and civil, leaping straight from the laboratory into passenger service (Owen 2001; Orlebar 1997). Loftin (1985, xi) captures the consensus view of aviation historians when he describes it as "an entirely different class of vehicle" to a traditional jetliner.

To an even greater degree than the A320 and B787, Concorde's radical design led aviation experts to doubt its safety. Issues like noise pollution and the effects of sonic booms tended to dominate public debates. Behind the scenes, however, were persistent concerns about the airframe's newness, the uncertainties that it introduced, and what those uncertainties could mean for the airplane's reliability. Observers on both sides of the Atlantic mobilized to voice doubts about the unforeseen hazards implicit in the radical departure from the traditional jetliner paradigm represented by Concorde (together with other putative supersonic passenger aircraft being designed at the time) (Wilson 1973). *The Economist* was one prominent voice; it mounted a sustained campaign against the airplane. The UK-based Anti-Concorde Project was another, claiming the support of thousands from academia and beyond.

Bo Lundberg, the director of a Swedish aeronautical research institute and a former test pilot, emerged as a particularly prominent skeptic. Writing in the *Bulletin of Atomic Scientists* in 1965, he outlined his safety concerns in a passage that neatly anticipates some core arguments of the preceding chapters regarding the limits of knowledge, innovation, and their relationship to reliability:

> [M]anufacturers have declared that supersonic airliners will be at least as safe as the jets today. The facts, however, suggest otherwise. An aircraft designer can normally do a great deal to minimize foreseeable risks. But he can do nothing about risks that he fails to foresee. The number of such risks increases rapidly with the number of new design and operational features that are introduced at one and the same time. In the past, a new type of civil aircraft has seldom incorporated more than one radically new feature, and this has usually been developed first on military aircraft. In contrast, supersonic transports cannot be built without introducing a host of radically new design features simultaneously, and relevant military experience simply does not exist. (Lundberg 1965, 29–30)

Lundberg makes a compelling point, and we must reckon with his concern if the argument about the importance of design stability is to stand. This, however, presents a problem, for as we have seen, Concorde is rarely, if ever, remembered as an unreliable design or unsafe airplane.

It is understandable that Concorde would be remembered more as being uneconomical than unsafe because, superficially at least, its safety record looks quite impressive. It carried passengers for over a quarter of a century with only a single catastrophic accident—a seemingly exemplary performance. And this service history presents a problem for the argument about design stability. The fact that the A320's reliability confounded its critics could be explained by pointing to its underappreciated design inheritance, but Concorde's service cannot be explained the same way. Concorde was an indisputably radical airframe, and insofar as design stability is vital to achieving the ultrahigh reliability expected of jetliners, then it should have fallen substantially short in this regard. If the argument of preceding chapters is to hold, its dissimilarity to other airframes should have introduced dangerous uncertainties and manifest as a deeply troubled service history punctuated by rational accidents. Why, then, is it not remembered as unreliable?

The solution to this puzzle is straightforward but counterintuitive. It is that, contrary to how Concorde is remembered, it actually *was* substantially less reliable than a traditional jetliner. Interpreted properly, its service history suggests that it paid a steep price in reliability for its glamorous novelty.

The "problem" of Concorde's performance, such as it is, arises because its unreliability is hidden by a paucity of data. To appreciate its true reliability—or lack thereof—it is necessary to think about that reliability statistically. The airplane enjoyed a long service life, this is true, but over the course of that life, it accrued a remarkably modest number of actual flight hours (BEA 2002, 145). Only twenty Concordes were ever built, six of which were prototypes that were exclusively used for testing. The fourteen that entered service, meanwhile, were in slower rotation than normal jetliners. As a result, they collectively accumulated only about 90,000 flights in their entire service life, far fewer than normal jetliners. (By comparison, the B737 made roughly 232 million flights over the same period, with significantly more hours per average flight, and the broader design paradigm that it represented made many times more.)

Seen in the context of this limited service, Concorde's single fatal accident statistically makes it the *least* reliable jetliner ever to enter mainstream passenger service in the West, and by a huge margin. If the B737 had been equally as reliable on a per-flight basis, it would have endured 2,578 accidents over the same period (and on a per-flight-*hour* basis, it would have endured considerably more again). Or compare it with the A320 family. As of late 2020, variants of the A320 had accrued fourteen fatal crashes over 119 million flights, amounting to 0.09 fatal accidents per million departures. Concorde's fatal crash-rate per million departures was 11.36, over 126 times higher, and with a 100 percent fatality record (*Airsafe* 2020).

Small numbers can be deceiving, and one accident could always be a statistical anomaly, but a closer look at Concorde's service history strongly suggests otherwise. Investigations subsequent to the Paris crash, revealed that the airplane had been suffering tire damage at an average rate of 1 per 1,500 cycles—many times higher than other long-haul jetliners. (Over the same period, for example, the A340 was losing tires at a rate of about 1 per 100,000 cycles [BEA 2002, 145–147].) With Concorde's limited service schedule, this translated into fifty-seven tire blowouts from 1976 to 2000, thirty-two of which damaged the airframe or engines in some way, including six instances where debris punctured one or more fuel tanks (Harriss 2001). In 1979, for instance, Air France Flight 054—en route to Paris from New York—was forced to divert to Dulles after passengers alerted flight attendants to a twelve square-foot hole in one of the wings. The airplane landed with one engine and two of three hydraulic systems inoperable, severed electrical cables, and fuel streaming from multiple tank penetrations (BEA 1980). It is highly plausible, therefore, that Jim Burnett, the chairman of the NTSB was correct when he commented about the Paris crash: "If there were as many Concordes flying as 737s, I suspect that we would have seen this kind of accident many times" (Harriss 2001).

The issues with tires and debris were far from isolated, moreover, as any detailed history of the airplane reveals myriad other problems and concerns. Throughout its life, Concorde was plagued by worrying incidents and problems. Its innovative composite rudders would sometimes delaminate or break apart midair, for instance, leading to two near-misses: one in 1989, the other in 1992 (AAIB 1989, 1993). (As a consequence, there were long periods where maintenance engineers deemed it necessary to perform detailed ultrasound scans after every fourth flight [Barker 2006].) As the airplane aged, operators repeatedly found cracks in its wings, and a 1998 study of its engines found so many endemic problems that it concluded that "[a] major technical event would probably end Concorde operation" (quoted in Harriss 2001).

Contrary to appearances, therefore, Concorde amply illustrates the importance of design stability and the risks of departing from it. Few expenses were spared in the airplane's development. It was tested far more extensively than any previous airframe—it was subject to nearly 5,000 hours of test flight, for example, over three times as many as the B747 (Donin 1976, 54)—and yet no amount of money or effort was able to recreate the reliability of a traditional jetliner.

(It is worth noting that this deficit is reflected also in the quality of the predictions that experts made about the airplane's performance: its radical novelty jeopardizing efforts to assess its reliability, as well as efforts to make it reliable. At the time of its certification, for example, the probability of a double tire-burst on takeoff was assessed to be on the order of one in every 10^7

flying hours. When Aérospatiale reassessed this probability in light of Concorde's first three-and-a-half years of service, however, it found that number might be off by two orders of magnitude [BEA 2002, 145]).

CONCORDSKI

Should any doubt remain about the dangers of innovation—or, indeed, the achievement represented by Concorde—then consider the Soviet Union's answer to the new airplane: the Tupolev Tu-144. The story of "Concordski," as Western commentators took to calling it, illustrates the same difficulties as Concorde, but much more starkly. Designed by the storied Tupolev design bureau, it entered service in November 1977 with superficially similar design and flight characteristics as Concorde—it flew at speeds of around Mach 1.6—but was plagued with problems from the start. Even when functioning as designed, the airplane had notable shortcomings. The cabin, for instance, is said to have been almost unbearably noisy, especially in the back, where, by some accounts, passengers sitting two seats apart were reduced to passing notes if they wished to communicate.

Yet the airplane rarely functioned as designed. From its very earliest days, the Tu-144 was extraordinarily unreliable—almost comically so when viewed from a distance. (The prototype has the dubious distinction of being the only jetliner to have ever been fitted with ejection seats.) Upon entering service, it manifested such a wide range of acute and persistent issues that Soviet officials eventually took the politically embarrassing step of reaching out to Concorde's Anglo-French manufacturers—first in 1977, and then again in late 1978—with requests for technical assistance. (Both requests were denied on "dual-use" grounds; Western authorities being concerned that any advice would have military as well as civil applications.) On the second such occasion, Tupolev provided a (probably bowdlerized) service record. It indicated more than 226 failures over a period of only 181 hours of operation, which had collectively caused the delay or cancellation of 80 out of 102 flights (Moon 1989, 197–199).

Designed for a service life of 30,000 flight-hours over fifteen years, the Tu-144 was prudently but ingloriously withdrawn from passenger service in June 1978 after seven months of service. Its tenure as a passenger aircraft had lasted a mere 55 flights, although it would limp on in cargo service until 1983, accruing a further 47 flights, for a grand total of 102. Over this period, it incurred two fatal crashes: the first during the 1973 Paris Air Show, and the second in May 1978, during a delivery test flight. Few technical endeavors speak more clearly to the importance of design stability to reliability.

For all this, however, the mere existence of Concorde and the Tu-144 raise another important question. This is because, even while the mixed records of these aircraft testify to the difficulties and dangers of radical innovation, the titanic expense and effort of creating them simultaneously testify to innovation's powerful allure. And, given that allure, it is reasonable to ask: why are these aircraft so anomalous? Or, put differently: why has the industry been so willing to commit to keeping the standard jetliner paradigm so stable for so long? As chapter 10 will explain, this is a surprisingly complicated question with significant ramifications.

10 SAFETY COSTS: THE STRUCTURAL FOUNDATIONS OF ULTRARELIABLE DESIGN

Of the major incentives to improve safety, by far the most compelling is that of economics.

—Jerome Lederer

10.1 AN ORGANIZATIONAL ANOMALY

GOOD BEHAVIOR

If "the story of Concorde was to demonstrate that the age of irrational decision-making was not yet past," as a magazine article put it (Gillman 1977), then the lacuna created by the airplane's retirement—the conspicuous absence of supersonic transport by the third decade of the twenty-first century—arguably speaks to an age of technological maturity, where airframers manage rational accidents by curbing their design ambitions.[1] The preceding chapters of this book have argued that the civil aviation industry transcends the uncertainties of its tests and models, in part, by committing to a common design paradigm and keeping that paradigm stable.

In helping resolve the epistemological problem of civil aviation's extreme reliability, however, the relative stability of its airframe designs poses an organizational problem. As we will see, stability for the airframers is expensive. It means forgoing, or long delaying, tempting innovations like composite materials that promise competitive advantages in a challenging marketplace. And while it is intuitive to see the industry's commitment to design stability as something driven by a concern for safety, organizations are generally thought to be poor at consistently prioritizing safety over profits.[2]

SAFETY SECOND

The finding that airframers are willing to consistently subordinate economic interests to safety is difficult to reconcile with the literature around organizations and technology. Outside of civil aviation, scholars routinely find that private companies are susceptible to moral hazards arising from incentives to maximize short-term economic gain (e.g., Power 1997), and organizations responsible for catastrophic technological systems are no exception to this rule.

Consider, just by way of illustration, the 2010 Deepwater Horizon disaster. The offshore drilling platform was exceptional even by the lofty standards of drilling platforms. Built by Hyundai at a cost of $350 million, the huge, semi-submersible rig was of a different class to most of its peers, holding records for the depths that it drilled (NCBP 2011: xiii; Transocean 2010). It met its catastrophic end in the Gulf of Mexico on April 20, 2010, while completing an exploratory well for the oil company BP. A blowout caused an explosion that killed 11 of its 126 workers and engulfed it in unquenchable flames. The stricken platform burned fiercely for two days before disappearing ignominiously beneath the waves. In its wake, it left a gushing oil leak at the seabed, 5,000 feet (1,500 meters) below, which took three months and billions of dollars to cap. To date, this environmentally devastating spill remains the largest ever in US waters, having contaminated 1,100 miles of coastline and 68,000 square miles of water (NCBP 2011).

As the scale of the disaster became evident, the White House convened a commission to investigate its causes. The commission's report, published the following year, attributed the accident largely to a corporate culture that encouraged workers to maximize revenue by cutting corners (NCBP 2011). It identified nine separate management decisions—each pertaining to the platform's design or operations—that increased the risk of a blowout, and it highlighted that seven of the nine unambiguously had saved BP money (NCBP 2011, 125). The specific contribution of each decision was ambiguous, the report concluded, but collectively they pointed to a common underling cause: a failure to consistently prioritize safety over short-term profits. Time and again, it found that "[d]ecisionmaking processes . . . did not adequately ensure that personnel fully considered the risks created by time- and money-saving decisions" (NCBP 2011, 125).

The findings of the Deepwater Commission might be shocking, but they should not be surprising. Indeed, any serious engagement with the literature around technological disasters—from academic ethnographies to formal accident reports—testifies to the regrettable normalcy of organizations resisting expenditures on marginal risk reduction (e.g., Vaughan 1999; Perrow 1999, 2015; LaPorte and Consolini 1991; Perin 2005; Hopkins 2010; Reason 1997; Silbey 2009; NCBP 2011; Sagan 1993). So abundant is this evidence that many social scientists consider it almost axiomatic that the safety of sociotechnical systems suffers when it clashes with profits, especially over time.

The difficulty of prioritizing safety over profit becomes more intuitive if we look at organizational incentive structures. Such structures and their relationship to risk behavior are always complex, as is the literature around them (see, e.g., Vaughan 1999; Gephart et al. 2009), but two generalizable observations are worth highlighting. The first is that organizations, and the individuals staffing them, invariably operate with regard to time frames that are ill suited to managing the kinds of high-consequence and low-probability risks that characterize catastrophic technologies. This is simply to say that—since management positions are only held for a few years at a time, and corporate strategies must cater to investors and quarterly earnings reports—it is difficult to create accountability structures that prioritize the prevention of very rare accidents, no matter how consequential those accidents might be. The second is that when technological accidents occur, the organizations held responsible for them rarely bear the full costs of those accidents. Insurance plays a role in this, as does the fact that courts and legislatures, conscious of jobs and shareholders, almost never penalize large corporations in ways that would undermine their viability.

The upshot of this, as Perrow (1994, 217) observes, is that the costs of accidents are rarely high enough to be more determinative of organizational behavior than wider economic incentives. The 2007–2008 financial crisis probably exemplifies this most clearly. Autopsies of the crisis and its causes routinely highlight the fact that banks rewarded traders—and investors rewarded banks—on the basis of short-term profits rather than long-term stability, and the price that these actors paid for the disaster was wholly eclipsed by the money that most of them accrued in making it possible (e.g., Lowenstein

2011). And although the financial crisis might be exemplary in this regard, it is far from unique. Studies of more classically "technological" accidents routinely come to similar conclusions. BP famously paid record penalties for the Deepwater Horizon incident, for instance, but those penalties still failed to cover the full cost of it. BP remained an extremely viable company (Uhlmann 2020), one that in 2018 reported an annual profit of over $49 billion.

The primacy that scholars ascribe to economic incentives becomes more intuitive if we try to appreciate the subtle ways that incentives can shape behaviors. Most accounts of economic pressures influencing organizational safety behaviors frame the process in terms suggesting moral compromise: deliberate corner-cutting, born of greed or irresponsibility. But such narratives are often misleading.

There is no shortage of greed or irresponsibility in the world, of course, but organizational decision-making is usually better understood in relation to the conditions that structure its reasoning. Consider, for instance, that the division of labor in bureaucracies invariably separates financial expertise from safety expertise, thereby creating circumstances where even well-intentioned economic decisions can have underappreciated safety ramifications. Consider also that the benefits of reliability engineering do not scale very intuitively with their costs. Achieving ultrahigh reliability requires organizations to spend large amounts of money to offset extraordinarily improbable failure modes: one-in-a-billion events that might never occur in the entire lifetime of the system. Such expenditures do not fit neatly into the kinds of economic calculus taught at business schools, and the divisions within bureaucracies help occlude their necessity.

"Safety" is an ambiguous condition, moreover, and, homilies notwithstanding, it can never be something that organizations prioritize absolutely. Even in optimal conditions, it is always being negotiated against rival considerations like cost and functionality. (The safest possible jetliner would be one that rarely left the ground and charged millions of dollars per flight, but few would consider this ideal.) No organization has unlimited resources to spend on risk reduction, in other words, and all must balance the safety of their operations against the ever-present risk of failing to compete (economically or otherwise): what Reason (1997) calls the tension between production and protection (see also Lazonick and O'Sullivan 2000; Froud et al. 2006; Perrow 2015; Bazerman and Watkins 2004, 5–7). Such considerations require complex and ambiguous judgments, and—when seen in light of the division of expertise, and the counterintuitive costs of reliability—it is easy to imagine how incentives might shape these judgments without presupposing much wrongdoing by the actors involved.

This is all to say that any organization that consistently prioritizes extreme, long-term safety over nearer-term profit—as civil airframers appear to do—deserves scrutiny. The organizations that make jetliners are clearly conscientious; they care about safety and employ many intelligent, earnest, and principled people. Any exposure to the industry speaks to this. But such considerations have limited explanatory value. It is reasonable to assume that most organizations responsible for catastrophic technologies care about safety and employ well-meaning people, but, as we have seen, most still fail to consistently prioritize safety.

There are few reasons to imagine that airframers should be exceptional in this regard. The economic pressures on them are intense. The market in which they operate is competitive and unforgiving; their customers have real options (or at least they have in the past—but this may be changing, as we will see in chapter 11); and the development costs of a new jetliner are such that they practically bet the company's future on the commercial success of new designs (Newhouse 1982; Mowery and Rosenberg 1981). (Lockheed, for example, was forced to exit the commercial jetliner market after its L-1011 Tristar failed to compete effectively with the DC-10.) And close accounts of airframers' circumstances, practices, and cultures, rarely suggest that they float free of the pressures that characterize other organizations (e.g., Schiavo 1997; Perrow 1999, 163–169; Fraher 2014; Barlay 1990; Oberstar and Mica 2008; Heimer 1980). Indeed, experts commonly voice concerns about the potentially corrupting effects of economic pressures on civil aviation's design choices (e.g., *Federal Register* 1998, 68642).

This is why the civil aviation industry's sustained commitment to design stability raises important questions with far-reaching implications for our understanding of organizations more broadly. Resolving these questions will take us on a somewhat circuitous—arguably even circular—journey. The excursion is worthwhile, however, because it sheds light on how the epistemology of ultrahigh reliability shapes the incentive structures around catastrophic technologies, and the role that regulators play in policing those technologies.

Let us start, therefore, by taking some time to further establish that the industry does indeed appear to be acting against its short-term economic incentives in pursuit of safety.

10.2 CHOOSING STABILITY

ECONOMIC SACRIFICE

The problem of why airframers pursue design stability would essentially disappear if stability itself was economically advantageous for them, and there is a credible argument that this is sometimes the case. Certainly, it is not always true that delaying or resisting innovation costs airframers money. Innovation is expensive, after all, and scholars often find that technological designs are kept stable by the costs of altering them (e.g., Musso 2009; Kemp, Schot, and Hoogma 1998; Freeman and Perez 1988; Mokyr 1990). This might be especially true in catastrophic technological contexts like civil aviation, where systems have extreme technical and regulatory requirements. Forgoing innovation in these contexts avoids the considerable development and testing costs that design changes incur, and it allows the industry to take advantage of time-honed efficiencies in manufacturing and servicing.

These savings can be significant, and they undoubtedly incentivize stability in many circumstances, but radical innovation is not always a poor investment, and it is easy to underestimate the pressures on airframers to innovate faster and more ambitiously. Among the many design changes that civil airframers abstained from, or long-delayed introducing, are some that seemingly promised to make their products meaningfully more competitive. Arguably, the clearest examples of these—let us call them "uncompetitive deferments"—involve innovations that promised to increase jetliners' efficiency.

To understand the pressures on airframers to increase the efficiency of their jetliners, and thus the costs of delaying or forgoing efficiency-maximizing innovations, it helps to consider the incentive structures that act on their primary customers: the airlines.

Airlines find it notoriously difficult to make money. The fact that they are highly regulated, together with the related fact that many are valued national assets, have historically limited their ability to consolidate, leading to chronic oversupply on many routes. The resulting competition—often distorted further by subsidies and, in recent decades, intensified by more transparent internet pricing—has squeezed revenues to a point where many airlines almost perennially hover on the edge of bankruptcy (Vasigh, Flemming, and Humphreys 2015; Gritta, Adrangi, and Davalos 2006). Over the period 1978–2010, the average net annual profit for the world's airlines was less than zero US dollars (–$0.04 billion, to be precise) (Cronrath 2017, 3). "If a capitalist had been present at Kitty Hawk back in the early 1900s, he should have shot Orville Wright," Warren Buffet told *The Telegraph* in 2002, adding, "He would have saved his progeny money" (Vasigh et al. 2015, 264).

In this environment of tight margins and tenuous profits, airlines are strongly incentivized to make hard-nosed economic calculations about airframe purchases. There are many factors at work in such calculations, but operating costs are the most prominent, and even small economies can make an enormous difference to an airline's bottom line. This is especially true in relation to fuel consumption, fuel being a major operating expense for most airlines and its price being their primary cost variable. (In 2012, for example, fuel alone was estimated to account for 33 percent of total airline operating costs [IATA 2013].)

The importance of fuel economy to airlines creates powerful incentives for competing airframers to adopt innovations that promise to improve the fuel efficiency of their jetliners. A design change that even marginally decreases a jetliner's fuel requirements would have to drive up its base price and maintenance costs by a lot before it became uncompetitive with less expensive but less efficient offerings from other manufacturers.

There are many ways of increasing the fuel efficiency of an airframe,[3] but the most straightforward of these is to make it lighter. The relationship between weight and fuel consumption is widely underappreciated. Heavier jetliners require more fuel to move—this is straightforward enough. But it is easy to forget that any additional fuel adds its own weight, which then requires more fuel, which adds more weight, which, at some point, requires structural reinforcements that add even more weight, which requires more fuel, and so on.

This irony of aeronautical engineering means that even small changes to the mass of an airframe can have disproportionate implications for an airline's operating expenses. Take, for instance, Southwest Airlines, which MIT researchers used as the basis for a 2013 study of the industry's fuel costs (Jensen and Yutko 2014). A smaller airline than many, Southwest operated about 1.6 million flights in 2013. According to the MIT model, if every passenger on each of those flights carried an extra cell phone, the combined weight of those phones would have cost the airline $1.2 million in fuel over the course of the year—a number would increase to $21.6 million if the passengers carried laptops instead (Jensen and Yutko 2014). (Understanding this relationship helps explain a lot about the behavior of airlines, from the growing trend of charging for baggage, to the decision by some carriers to stop offering free bottles of water. It is partly why American Airlines replaced pilots' emergency binders with iPads—a measure that it anticipated would save it $1.2 million per year—and why All Nippon Airways started requesting that passengers visit the bathroom before boarding [Brown 2009; Jensen and Yutko 2014]).

Such figures offer an important perspective on civil aviation's adoption of innovations like fly-by-wire and advanced composite materials. Both these adaptations promised to make airframes significantly more efficient, primarily by reducing weight. And, once embraced, they lived up to this promise. The fly-by-wire A320 was claimed to be 40 percent more fuel efficient than the B727, with which it was designed to compete (Beatson 1989). The composite B787, meanwhile, was designed to consume a fifth less fuel per passenger mile than the B767 it was intended to replace (Marsh 2009, 16–17; Waltz 2006). These exact figures are contestable, and in each case there were multiple factors that contributed to the airframes' efficiency gains, but both aircraft were undoubtedly more efficient, and fly-by-wire and composites played important roles in these achievements. Even though both innovations were costly to implement, therefore, they promised significant efficiencies for customers that had averaged negative returns over most of their history.[4]

For airframers eager to offer competitive products, the efficiencies offered by fly-by-wire and composites—both clearly viable innovations that had long been employed in other aviation contexts—must have loomed large against abstract and invisible misgivings about reliability. It would have been economically rational for executives to push the earlier adoption of both in efforts to sell more jetliners than their rivals. At least in respect to these design choices, therefore, it is fair to say that keeping airframe architectures stable—changing them very slowly and long delaying innovations until they had been validated exhaustively in other spheres—came at a cost. It is difficult to interpret this behavior except in terms of a consistent ability to prioritize long-term safety objectives over shorter-term economic incentives.

Whether this ability reflects any real restraint on behalf of the airframers is another question, however, as it is probably more intuitive to imagine design stability being imposed on airframers by regulators. Airframers are kept from innovating too aggressively, we might suppose, by the FAA and its elaborate code of certification requirements. After all, policing the industry is ostensibly one of aviation regulators' primary functions.

As we will see, however, this too is problematic.

10.3 REGULATING STABILITY

AN INTUITIVE NOTION

The idea of aviation regulators policing the stability of jetliner architectures is undoubtedly appealing. Preceding chapters of this book outlined in detail why formal rules and metrics, with their many ambiguities and interpretive flexibilities, do not allow experts to monitor and enforce ultrahigh reliabilities directly. (As we have seen, a jetliner could meet every standard, pass every test, and still be unsafe to fly.) Nevertheless, it might still be possible for regulators to monitor and enforce practices like design stability. Innovation scholars sometimes speak of "technology-forcing" regulations, designed to incentivize actors to take design risks in pursuit of newer and better technologies (e.g., Gerard and Lave 2005; Lee et al. 2010). In the context of catastrophic technologies, by contrast, it arguably makes sense to think of "technology-halting" regulations, intended to promote design stability in pursuit of safety.

There is evidence to support this idea. Type certification is ostensibly framed around quantitative reliability targets as a way of avoiding innovation-constraining design stipulations, but it clearly rewards design stability. If the regulator finds a proposed system to have insignificant differences compared to a previously certified design, for instance, then the rules largely (or wholly) allow it to retain its original certification. If it deems a system to be unusually innovative, by contrast, the rules mandate extra scrutiny, often requiring the formulation of an "issue paper" to identify concerns and establish additional assessment protocols (FAA 1982, appx 1; NTSB 2006b). The effect of such measures is to make it significantly easier and cheaper for manufacturers to gain approval for systems that are identical to, or strongly derived from, previously certified designs.

Still, however, direct regulation offers an unsatisfying explanation for the industry's design stability.

DIRECT OVERSIGHT

To understand the FAA's ability (or rather their inability) to police the stability of jetliner architectures, it helps to think about the fundamental nature of the task. Innovativeness is a fundamentally ambiguous and interpretive property of artifacts that, like reliability itself, is impossible to formalize or objectively quantify. Assessing or controlling the stability of a system would require regulators to assess its similarity to previous systems. As we have seen, however, the question of when two things are the same—be they redundant elements, or test and real-world environments—raises extremely complex relevance questions. And navigating such questions necessarily involves a great number of consequential but qualitative technical judgments, which regulators are ill equipped to make.

There are several reasons to doubt the FAA's ability to directly make the kinds of judgments required to effectively police the innovativeness of new systems. The most straightforward is simply that it lacks the necessary resources to grapple with the nuances of new systems. It is difficult to say with precision how many technical people the FAA employs to assist directly in certification activities, but as of 2006 it was 250 (NTSB 2006b, 68), and the agency has not transformed dramatically since then. Whatever the exact number, however, there is no question that it is far fewer than would be needed to actively police the innovativeness of every system in a new airframe, especially given the industry's increasing reliance on subcontractors (FAA 2008a; Bonnín Roca et al. 2017).[5] In 2019, the FAA's acting administrator told Congress that his agency would need at least 10,000 additional employees to directly perform all its certification duties (Shepardson 2019).

Even if afforded infinite resources, moreover, there is good reason to believe that the FAA would lack the technical intimacy needed to make the requisite technical judgments. A range of official analyses have testified to this deficit. As early as 1988, for example, the Office of Technology Assessment (OTA) was reporting that FAA regulators lacked the expertise to make sound technological judgments (OTA 1988), a conclusion that the US Government Accountability Office (GAO) echoed in 1993, when it found the agency to be "not sufficiently familiar with [specific systems] to provide meaningful inputs to the testing requirements or to verify compliance with regulatory standards" (GAO 1993, 19). The reports' findings echo those of the Deepwater Commission, which similarly found the platform's regulator lacked sufficient personnel and expertise to keep pace with the industry's technological developments (NCBP 2011, 56–74). They also echo many academic studies of technological practice, which routinely find that the understanding needed to navigate complex technical judgments is born as much from close familiarity as from formal analysis or abstract study (e.g., Collins 1982, 1985, 2001, 2010; MacKenzie and Spinardi 1996). Put in STS terms, we might say that there is an inevitable imbalance—or what Spinardi (2019) calls an "asymmetry"—between the expertise of regulators and that of airframers, wherein the former lack the "tacit knowledge"[6] to make the judgments on which key questions about design stability would hinge (Downer 2009a; 2010).

For both these reasons—a deficit of resources and of intimacy—the idea of external regulators effectively policing airframers' technical judgments to enforce appropriate levels of design stability is almost certainly unrealistic. As one aeronautical engineer succinctly put it, "[T]here is not a way for a third-party organization to assess our understanding of [complex avionics] . . . The very best method we have of discriminating between those who can, and those who can't but talk a good game, are their peers."[7]

The FAA actually concurs. This is why it has a longstanding practice of deputizing engineers from within the industry to act as its surrogates and make technical determinations on its behalf. Herein, therefore, lies a more subtle mechanism by which the regulator might be policing the stability of civil airframes: not directly, via assessments of the technology itself, but indirectly, via assessments of the personnel and organizations that design it.

As we will see, this mechanism is also insufficient for policing design stability, but it is worth exploring nevertheless.

SECOND-ORDER OVERSIGHT

The FAA's tradition of deputizing engineers to act on its behalf has roots in the earliest days of certification. "We've got certain safety factors, and we'll have our engineers check your plans with respect to them," William Mac-Cracken, the man charged with framing the first regulations, told manufacturers, "but mainly we'll rely on you to comply voluntarily" (quoted in Komons 1978, 98). Today, this relationship is formalized in what the FAA calls the "designee program." The program employs a variety of designees across various roles (FAA 2005b), but the group that plays the most prominent role in its assessment efforts is referred to as "Designated Engineering Representatives (DERs)" (FAA 2005a).

In many ways, DERs are the backbone of the certification process, overseeing tests, calculations, and designs to ensure that systems are compliant with regulations. In this capacity, they answer to the FAA, but most are also employees of the manufacturers (although a small number of consultant DERs work for third parties or independently). Most have about fifteen to twenty years of experience and hold key technical positions working on the systems they assess.

The responsibilities afforded to designees have expanded considerably over time. When the program began, it was intended that they would conduct well-defined tasks, allowing regulators to concentrate on larger oversight functions: framing requirements and analytical criteria, designing tests, and making final compliance determinations. As jetliners became more complex, however, the roles originally demarcated for regulators became increasingly untenable. As a result, DERs may now be authorized to assume almost all key oversight functions, to the point where it is now common for them to both frame and witness tests on the regulator's behalf (FAA 2007, 44; GAO 2004; NAS 1980).

Periodic investigations of the certification process illustrate this growth of DER responsibilities. In 1989, for instance, an internal FAA review concluded that the regulator had been forced to delegate practically all the certification work on the 747–400's new flight-management system because its staff "were not sufficiently familiar with the system to provide meaningful inputs to the testing requirements or to verify compliance with the regulatory standards" (AIAA 1989, 49). Four years later, the GAO reported that the FAA was increasingly relinquishing roles that it traditionally retained, concluding that between 90 and 95 percent of all regulatory activities were being delegated, including many "core" functions such as the framing of standards (GAO 1993, 17–22). In keeping with this shift, the ratio of designees to core regulatory personnel has changed dramatically since the 1970s. Between 1980 and 1992, for instance, the number of designees overseen by the FAA's two main branches rose 330 percent, while core regulatory personnel rose only 31 percent, bringing the overall ratio of designees to regulators from about 3-to-1 in March 1980 to 11-to-1 in 1992. (GAO 1993, 17–19). This trend has continued since the 1990s. By 2006, for example, the NTSB (2006b, 68) was reporting that FAA Certification Offices employed 250 personnel while drawing on 4,600 designees: a ratio of over 18-to-1.

The designees offer the FAA a level of hands-on tacit knowledge and technical intimacy that its regulators, as outside observers, cannot match (NAS 1980, 7; Fanfalone 2003). Significantly, they also give its regulators access to civil aviation's social economy and reputational landscape: its rumors, hearsay, and culture. This is important because it means that even if regulators struggle to judge the significance of innovations directly, they are still likely to notice if engineers who can make those judgements have significant misgivings about a change, even amid the background noise of normal engineering dialogue and dissent. Construed in STS terms, we might say that the FAA uses the designee system to access what Collins (1981, 1985, 1988) would call the "core set" of aviation engineering: the narrow community of informed specialists who actively participate in the resolution of technical controversies. Regulators might lack the expertise required to actively participate in technical debates—what Collins and Evans (2002) call "contributory expertise"—but possess the competence necessary to understand what it means to be an expert participant, as well as the familiarity required to engage with such experts ("referred" and "interactional" expertise, in Collins and Evans's terminology). All the regulator has to do in these circumstances is assess the creditworthiness of the designees themselves.

Understood in this way, we might say that the DER program helps the FAA manage the unavoidable limitations of its knowledge by substituting an intractable technical problem for a much more tractable social one. Recognizing its limited ability to make complex engineering judgments directly, the FAA instead judges the actors who are capable of making technical judgments: a process that I have elsewhere referred to as "second-order" oversight (Downer 2010). The idea of the FAA actively policing design stability seems more plausible when construed this way. To the extent that regulators can draw on the insight of their designees, then they might be well positioned to make effective technical rulings about the significance of different design changes.

It is worth noting that external reviewers of the FAA's practices have endorsed its delegation program on these grounds (e.g., NAS 1998). It is also worth noting that many other technological domains have come to similar arrangements. It is actually very common for organizations producing high-risk technologies to play an active role in their own regulation, "if only because they alone possess sufficient technical knowledge to do so," as Perrow (1984, 267) puts it. Nuclear regulation operates on similar principles, for instance, as does railway regulation (Perin 2005; Hutter 2001).

While second-order oversight probably does serve a valuable function, however, it still offers an unsatisfactory explanation for the industry's enduring ability to subordinate profits to safety. This is because the FAA and its delegates are almost certainly too close to the industry that they regulate to be expected to steer it, effectively and consistently, in directions that it does not wish to travel. To invoke a slightly loaded and misleading term, we might say that the FAA and its designees are both highly vulnerable to "regulatory capture."

CAPTURE

First gaining traction in the 1970s (e.g., Peltzman 1976; Posner 1971, 1974, 1975; Stigler 1971), the concept of "regulatory capture" describes a process whereby powerful institutional actors come to control, influence, or otherwise dominate the bodies charged with regulating them. Capture "puts the gamekeeper in league with the poacher," so to speak, by creating circumstances where organizations can pursue their self-interest in ways that regulators are expected to curb on the public's behalf; or even, on occasion, circumstances where they can use regulation to further their own ends at the public's cost (Wiley 1986, 713).

Critics of the FAA frequently describe it as being captured by the airframers that it is supposed to police (e.g., Stimpson and McCabe 2008; Dana and Koniak 1999; Niles 2002; Nader and Smith 1994; Schiavo 1997; Perrow 1999; Fraher 2014; Oberstar and Mica 2008). Such assertions usually point to entwined interests and sympathies operating at varying levels of abstraction (e.g., Bó 2006). On the level of individual sympathies, for instance, critics highlight the fact that FAA personnel often remain in specific regional offices for years at a time and often become close with the people they oversee. This is by design. Fostering the kinds of interactional expertise outlined here requires that regulators develop close, long-term working relationships with their charges. But effective regulation is traditionally thought to hinge on the maintenance of emotional distance between agents of each party, and regulators in many other spheres are regularly rotated through positions and locations to mitigate capture risks.

On a different level of abstraction, critics often point out that aviation regulators and airframers share broad, structural-level interests and incentives. The FAA is a US government agency that certifies jetliners built by US corporations. To date, this primarily means Boeing, a significant national economic asset, and a major defense contractor with a powerful lobby in Washington and beyond. It does not require much cynicism to imagine that such considerations have consequences, especially insofar as they pertain to questions that might jeopardize the company's future, as the viability of a new jetliner might. (It is also worth noting in this context that the FAA was originally established to both regulate and promote the industry. It held this dual mandate until 1996, when it was amended in the wake of a high-profile accident, via legislation that changed the regulator's mission from "promoting" to "encouraging" the industry, without requiring any changes to its "organization or functions" [Mihm 2019]).

To understand these relationships and the problem that they pose, it is important to remember that incentives can shape judgments without implying moral failings. "Capture" is a loaded term. It is often portrayed as an almost conspiratorial process wherein regulators knowingly subvert clearly defined rules. Such associations often misconstrue the nature of the problem, however, and they run contrary to any real exposure to regulatory personnel, who invariably appear professional and sincere. Rather than seeing capture as fundamentally conspiratorial, it is more productive in this context to think of it as a subtle process wherein aligned sympathies and shared worldviews come to colonize irrevocably ambiguous rules and interpretations—a dynamic less analogous to a poacher and gamekeeper, we might say, than to an honest figure-skating judge who has a close relationship with the skater being judged.

This is to say that capture needs to be understood in relation to the interpretive flexibility of technical decision-making, wherein crucial regulatory questions hinge on subjective judgments about everything from the representativeness of artificial birds to the similarity of redundant hydraulic systems. There is extensive evidence that subjective judgments inevitably come to be shaped by structural interests. Scholars since Karl Marx have argued that interests tend to invisibly colonize cultures and interpretations, a finding supported by modern psychology (e.g., Kahneman 2011), by historical evidence that engineers routinely underestimate risks of their own designs

(Petroski 1992a), and by the STS literature, which has long held that interests and predispositions permeate even the most rigorous knowledge claims (e.g., Bloor 1976). To imagine that FAA regulatory determinations were an exception to this rule would be to defy generations of social research. It would also go against explicit concerns expressed by secretary of transportation Mary Peters (in testimony to Congress), and by the FAA itself (in internal memos and other semiprivate communications), about a culture of excessive "coziness" between regulators and airframers (e.g., in Oberstar and Mica 2008, 12; Mihm 2019).

None of this should be read as suggesting that regulators play *no* role in shaping new airframes and promoting reliable design. Regulators are in constant dialogue with airframers, which do not always get their way. (When Niles [2002, 384] quotes an FAA veteran as saying: "To tell the truth, the industry, they really own the FAA," that is probably an exaggeration.) Neither should it be read as suggesting that the FAA is exceptional in being captured by the industry that it regulates. Its relationship with airframers is typical of catastrophic technology regulators more broadly (see, e.g., Perrow 2015).[8]

It is important to recognize, however, that the regulator's abilities and agency are limited, and that, insofar as civil aviation exhibits an extraordinary organizational commitment to safety, regulatory oversight is not a credible explanation for that commitment. Indeed, a senior FAA official conceded as much at the Flight Safety Foundation's 1990 annual International Air Safety Seminar. "The FAA does not and cannot serve as a guarantor of aviation safety," he told the audience. "The responsibility for safe design, operation and maintenance rests primarily and ultimately with each manufacturer and each airline" (Nader and Smith 1994, 157).[9]

To return to the central theme of this chapter, this still leaves us with a problem regarding the behavior of the aviation industry. Airframers seem to be making choices that consistently prioritize long-term reliability over short-term economic incentives. This behavior runs contrary to widely held sociological expectations about the nature of organizations, and it seems unlikely that regulation could be its primary explanation. What is it, then, that makes civil aviation different?

The answer, I suggest, lies in a factor that makes the industry exceptional in so many other ways: the unique volume at which it operates.

10.4 COSTS REVISITED

ACCIDENTS AND INCENTIVES

At the start of this chapter, I argued that most organizations are incentivized to underinvest in avoiding catastrophic technological accidents because they rarely bear the full economic costs of those accidents, and because they operate on time frames that are difficult to reconcile with extremely infrequent events. Some of these conditions apply very clearly to civil aviation. Airframers are rarely penalized directly for design-based technological failures, for example, not least because it is difficult to sue them for designs the FAA has formally certified as reliable. (For its part, the FAA itself cannot be sued for perceived shortfalls in its oversight practices [see Lagoni 2007, 247].) At the same time, however, there are good reasons to imagine that civil aviation is exposed to accidents in a way that sets it apart from other catastrophic-technological spheres.

Simply put, civil aviation—compared to other catastrophic technological spheres—might be expected to have a unique relationship to accidents for the same reason that it has a unique relationship to epistemology: its service experience. As outlined in chapter 3's discussion of the aviation paradox, civil aviation works on a radically different scale than its technological peers. With tens of thousands of jetliners operating simultaneously, they accrue service hours much faster than any other catastrophic technology. This service experience has many ramifications regarding things like recursive practice, as we have seen, but one hitherto unexplored consequence lies in the way it shapes the industry's structural incentives.

If every jetliner sold in the same numbers and operated on the same schedule as Concorde, then, practically speaking, it might be cost rational for the industry to compromise on reliability by embracing innovations that promised to make their products more competitive in the near term. In this scenario, the infrequently flying jetliners would accrue service slowly, so their designs could be much less reliable than modern airframes and decades might still pass before a manufacturer saw its first catastrophic accident (or certainly before any generalizable design shortcomings became statistically demonstrable). By that time, the individuals who designed, assessed, and approved the fallen jetliner would likely have retired or moved on in their careers. The airframer, in turn, would likely be selling a substantially changed product, which it could plausibly claim to be free of the same short-comings. If history is any guide, moreover, then the fines that it incurred from the accident would not be especially punitive. The same goes for any reputational damage, which would likely benefit from narratives—in newspapers, legislatures, and courts—that contextualized the accident in relation to the jetliner's preceding decades of safe service and its differences from current offerings.

Most jetliners sell in much greater numbers than Concorde, however, and operate on much more demanding schedules. This changes everything, as it means that they accrue service hours at rates that reveal any reliability shortcomings quickly. With a huge number of flights every year, it is likely that whatever can go wrong with a jetliner, will, and in a time frame that affects airframers in their present form, managers in their current positions, and products in their current incarnation.

In these circumstances, therefore, catastrophic failures have real and direct consequences, both for the airframers as organizations and for the individuals who staff them. For the airframers, these consequences can be financially devastating, not because of fines or penalties, but because the market in which they operate is unusually sensitive to public confidence, and public confidence is sensitive to the absolute frequency of accidents. As outlined previously, airframers operate in a competitive market where sales hinge on the decisions of customers (the airlines), with real choices and fickle customers of their own. In these circumstances, unreliability, or even the appearance of it, can have severe financial ramifications regardless of any formal liabilities (e.g., Cobb and Primo 2003, 5).

The conditions described here radically shape the incentives around reliability. If a jetliner is less than ultrareliable by design, then—because of the industry's high operating volume—this is likely to manifest in accidents before its airframer has stopped taking orders for it. If those accidents give that jetliner (or, almost as likely, its manufacturer) a poor reputation, passengers will start avoiding certain aircraft when booking flights, which will put pressure on the airlines to cancel orders for those aircraft and to look elsewhere for future purchases.

The reality and significance of such confidence spirals in civil aviation are amply illustrated by the industry's history, which is punctuated with once-illustrious manufacturers brought low by tarnished safety reputations. Take, for example, the two 1954 Comet crashes discussed in chapter 7. De

Havilland revised the Comet's airframe in the wake of the accidents' investigation, resolving the underlying fatigue issue. And, following this work, there were no principled reasons to suspect that the Comet would be any less reliable than its peers (military variants of the airframe would go on to serve for six decades). Commercially, however, neither airframe nor airframer would ever recover. The accidents irreparably damaged the Comet's reputation, effectively ending the pioneering British manufacturer's bid to be a major player in the emerging industry.

In many ways, De Havilland's collapse was the opening act of the jet age, and the industry took heed. Perhaps even more salutary, however, was McDonnell Douglas's experience with the DC-10 almost two decades later.

The DC-10—an impressive airplane in many ways, but one that was famously designed to a tight budget—was born with a design weakness. It involved the door to the cargo hold, and, we might reasonably say, probably owed more to economically driven compromises than to the inherent epistemological ambiguities of complex systems. The cargo doors to most jetliners open inward. This makes them more secure, as they can be shaped to act like a plug when the airplane is pressurized from the inside. Inward-opening doors take up valuable cargo space, however, so the DC-10 was designed to open outward. This decision had the effect of making the door's locking mechanism safety critical, but its designers did little to treat it as such. The mechanism lacked redundancy and was designed in a way that sometimes allowed it to appear securely locked when closed improperly, and even register as such in the cockpit.

The dangers of this design became apparent in June 1972—just nine months after the DC-10 entered service—when the cargo door of American Airlines Flight 96 blew out shortly after it took off from Detroit (NTSB 1973). The blowout caused the floor of the jetliner to partially collapse into the cargo bay below, but the pilots were able to land without casualties. It was a close call, however, and per the NTSB's recommendations, McDonnell Douglas redesigned the locking mechanism and added vents to keep the cargo bay from decompressing explosively. Neither airframer nor regulator mandated changes to aircraft already in service, however, and the redesigns proved insufficient to remedy the underlying problem (AAIB 1976). This became tragically evident two years later, in March 1974, when an almost identical failure struck Turkish Airlines Flight 981, shortly after it left Paris. On this occasion, the aircraft fared less well. The floor collapsed in a way that damaged its control cables, and it crashed into a forest with the loss of all 346 on board (AAIB 1976).

In the DC-10, we might say, McDonnell Douglas committed two meaningful transgressions. The first was to prioritize the DC-10's competitiveness over its safety by breaking with the standard airframe paradigm in a way that traded technical certainty for economic advantage. The second, more cardinal, sin was a lapse of recursive practice: it failed to fully grasp the nettle of its error, and, as a result, lost a second aircraft to the same cause. The airframer paid a steep price for these lapses. It quickly remedied the issue after Flight 981—the DC-10 has never failed in the same way since—but the damage was done. The controversy weighed heavily on public perceptions of the DC-10, which gained a reputation for unreliability. This reputation was arguably unjust—as of 2008, the DC-10's lifetime safety record was comparable to other jetliners of its generation (Boeing 2010; Endres 1998, 109)—but it shaped the coverage of later incidents,[10] and travelers took notice. Bookings on the DC-10 began to suffer, and the airlines responded. TWA put out full-page advertisements to reassure potential customers that it owned none of the aircraft. American Airlines could not make the same claim, but it ran campaigns stressing that it serviced certain routes exclusively with Boeings. Embroiled in controversy, the DC-10 tanked as a marketable product, with airlines across the world canceling their purchase options. McDonnell Douglas might have weathered the storm, but the airplane's tattered reputation became a millstone. It dropped the "DC" designation for its next jetliner—the MD-11—in an effort to distance it from its predecessor, but it failed to sell, despite being highly regarded by engineers. Saddled with debt from the airplane's development, the historied McDonnell Douglas Corporation faltered and was eventually subsumed by Boeing (Newhouse 1982; Davies and Birtles 1999; Waddington 2000; Endres 1998).

SELF-INTEREST

Herein, therefore, lies a plausible solution to the puzzling sociology of civil aviation's design choices: it has unusual incentives, which happen to align with reliability, rather than an unusual relationship *to* those incentives. In an industry that is so aware of its own history, it would be strange indeed if the fates of the Comet and DC-10—and those of their manufacturers—did not weigh heavily on the decision-making of other airframers, pulling against shorter-term incentives to maximize profit by innovating in ways that compromised reliability. Or, at the very least, we can imagine that these fates acted as a form of "natural selection," wherein only airframers that prioritized reliability (and thus design stability) survived for long.

A succession of formal investigations into the FAA have come to an equivalent conclusion: repeatedly finding that concerns about its regulation practices—regarding capture or delegation, for example—are moot because of the industry's financial incentives keep manufacturers in line. As early as 1980, for instance, the National Academy of Sciences (NAS) was highlighting the importance of the self-interest of the manufacturer in designing a safe, reliable aircraft that would not expose them to lost sales and litigation from high profile failures (NAS 1980); a view the GAO (2004) echoed over twenty years later.

As we have seen at length, the choices airframers have made also seem to bear this out. When it comes to the reliability of airframes at least, it is difficult to fault them. The last half-century of civil aviation produced many jetliner accidents, simply as a function of its enormous volume, but it is extremely rare that jetliners have failed in ways that might be attributed to an overeagerness to innovate.

Rare does not mean unheard of, however, and somewhat inconveniently for the argument in this chapter, two Boeing 737-MAXs crashed during its drafting, embroiling the company in a crisis seemingly born of ill-managed innovation and short-term economic pressure. Chapter 11 will address that crisis, its causes, and its implications. It will then explore the industry's relationship to a property known as "crash survivability," which, I will argue, offers further evidence that economic interests, more than oversight or good intentions, drive jetliner design choices.

11 INCENTIVES IN ACTION: ON DEFICIENT 737s AND NEGLECTED SURVIVABILITY

When it comes to profit over safety, profit usually wins.
—James Frazee

11.1 BAD PRACTICES?

A CONTROVERSIAL NOTION

Type "civil aviation news" into your preferred search engine and, on most days at least, the results will amply testify to the fact that the industry's safety practices are far from being above reproach. Not all reporting is well informed, of course, and not all reproach is fair, but US civil aviation has no shortage of credible critics. Ralph Nader, the crusading safety advocate and four-time presidential candidate, for example, has painted the industry as deficient in its commitment to safety (Nader and Smith 1994; Bruce and Draper 1970). So has Mary Schiavo, a former inspector general of the US Department of Transportation, who all but dedicated her retirement to excoriating the FAA for what she sees as dangerous negligence (Schiavo 1997). Such criticism poses a problem for the argument of chapter 10, in that it seems to contradict the claim that structural incentives lead airframers to prioritize long-term safety of their own accord. As with the claim about design stability, therefore, it is again worth pausing, briefly, to explore and substantiate this claim in more detail before moving forward with the wider argument.

To this end, this chapter will discuss two prominent test cases, both of which, in different ways, appear to challenge the argument that the industry is driven by economic incentives to pursue safety. Section 11.2 will look at the manifest failings around the Boeing 737-MAX; examining what the accidents suggest about the airframer's choices. The MAX's shortcomings clearly illustrate the powerlessness of regulators, I will argue, but they have more ambiguous implications for the industry's relationship to economic incentives. Section 11.3 will look at the industry's troubled relationship with crash-survivability measures, which are intended to make accidents less fatal. Interpreted properly, I will argue, the relationship offers much more straightforward and compelling evidence that economic incentives have shaped the industry's safety choices. As with the MAX, moreover, the story of crash survivability speaks to the difficulties that regulators face in policing jetliner designs, evidencing both the fact of those difficulties and their causes.

Much as the stories of Concorde and advanced composites in chapter 8 helped illustrate and substantiate the industry's commitment to keeping airframe designs stable, so the stories that follow shed light on its willingness (and the limits of that willingness) to prioritize long-term reliability over short-term profits.

Let us turn first, then, to the 737-MAX.

11.2 COMPROMISED BY DESIGN

TWO DISASTERS

In October 2018, thirteen minutes after leaving Jakarta, LionAir Flight 610, a new variant of Boeing's venerable 737—the 737-MAX—began climbing and diving erratically until it eventually plummeted into the Java Sea, with the loss of all 189 passengers and crew. A preliminary report into the accident by Indonesian authorities pointed to mechanical issues with the airframe, but Boeing, which had introduced the MAX only the previous year, vigorously contested this finding. It offered a competing account that exculpated the airframe and focused instead on factors relating to maintenance and airline operations. Boeing's account assuaged the fears of global aviation authorities, which allowed the MAX to continue operating unhindered. At least, they did until almost five months later, in March 2019, when a second MAX—Ethiopian Airlines Flight 302—crashed under eerily similar circumstances. Soon after leaving Addis Ababa, the plane's altitude began fluctuating wildly until, six minutes after takeoff, it struck the ground at almost 700 miles per hour, again with the loss of all on board (Nicas and Creswell 2019; KNKT

2018). Global aviation authorities grounded the MAX after the second accident, and in the months that followed, there emerged a broad consensus that the airframe was at fault after all.

The primary cause of both accidents was determined to be the airplane's computerized flight controls. More specifically, the crashes were found to have been instigated by a system called the Maneuvering Characteristics Augmentation System—invariably referred to as "MCAS."

Boeing had developed MCAS specifically for the MAX in order to alter the airplane's flight behavior. As a variant of the 737, the company had designed the MAX to be a near-replica of the original, type-certified incarnation of that airframe: identical in every way except in the narrow areas where it was intended to differ. (The hardware running its flight software was decades old, for instance, and it even eschewed fly-by-wire to retain the original's hydraulic cables [Nicas and Creswell 2019; Ostrower 2011; Nicas et al. 2019]). But one way in which the MAX was intended to differ from preceding 737 variants was that it would employ larger, high-bypass engines. Earlier 737 variants had been too low to accommodate newer, more efficient, engines, so Boeing had made room for them on the MAX by slightly lengthening the front landing gear and, importantly, by moving the engines forward on the wing.

Moving the engines forward, however, had complex ramifications for the airplane and its economics. It altered the airplane's flight characteristics, giving it a tendency to pitch up more when climbing. This was not necessarily hazardous in itself, but it threatened Boeing's characterization of the MAX as a changed version of the already-certified 737, which created a problem. If the MAX flew differently from the rest of the 737 family, then there was a danger that regulators would subject it to significantly greater oversight requirements and demand that its pilots be extensively retrained, both of which would raise the MAX's costs and thereby lower its competitiveness. Hence MCAS: a suite of digital flight controls that were intended to make the MAX the same as the rest of the 737 family, but inadvertently created catastrophic differences.

The two MAX accidents were caused when faulty—and, notably, not redundant[1]—Angle of Attack (AOA) indicators communicated the wrong pitch information to MCAS. In both cases, this led the computer to keep pushing the airplane into a dive, repeatedly overriding its pilots, who struggled against their aircraft in confusion, and ultimately in vain (Nicas and Creswell 2019; KNKT 2018).

EXEMPLARY FAILURES

In many ways, the MAX disasters illustrate the arguments and themes outlined in previous chapters. They exemplify the importance of redundancy, for example, and the complex interpretations involved in implementing it. (The decision not to make the AOA indicators redundant, for example, was justified on the judgment that the pilots could be considered an acceptable backup [Nicas et al. 2019; Nicas and Creswell 2019]). They also exemplify the difficulties that arise from the problem of relevance, as well as how those difficulties can contribute to accidents. (The flight simulators used for training MAX pilots failed to adequately reproduce MCAS behavior, for instance, and flight tests of the airframe were conducted within a normal operating envelope, which did not reproduce the AOA failure [Nicas et al. 2019].) More broadly, they exemplify the difficultly of judging the significance of design changes when making safety determinations, as well as the powerlessness—or perhaps unwillingness—of regulators to effectively police those determinations (Nicas et al. 2019; Tkacik 2019; Mihm 2019; Robison and Newkirk 2019). For whatever reason—be it captured regulators, the limitations of tests, or something else—a dangerous design flaw evaded the certification process.

Beyond this, the MAX disasters exemplify the dangers of innovation in civil aviation and the costs that can accrue to airframers that pursue it incautiously. For the first time in decades, two jetliners failed catastrophically because of a common design weakness that arguably might have been avoidable, or, at minimum, should have been correctable after the first failure. As such, Boeing committed the cardinal sin of civil aviation, and it is paying a steep price for its transgression.

It will probably be a long time before we can say with any authority how much the crisis will ultimately cost the company, but it is already an onerous price. Shortly after the second accident, the MAX was grounded by regulators across the world, instantly shuttering almost 400 recently delivered jetliners until they could be reeingineered and recertified (which the FAA eventually approved in November 2020, and EASA approved in January 2021). This had far-reaching financial repercussions for Boeing (Kitroefff and Gelles 2019). In January 2020, the company reported that the crisis would cost it at least $18.7 billion, a figure that included $8.3 billion in compensation to airlines and $6.3 billion in increased manufacturing costs (Isidore 2020b). This was a best-case scenario, however, and was premised on the airplane returning to operation in the middle of that year, which did not happen. That figure also excluded litigation costs, the scope of which became apparent in January 2021, when Boeing agreed to pay an additional $2.5 billion in penalties and damages after being charged with fraud for concealing information from safety regulators (DoJ 2021).

As with the Comet and DC-10, moreover, the most serious financial repercussions arising from the MAX are likely to come long after the accidents, in the form of reputational damage and lost sales. Early into the crisis Boeing launched a major publicity offensive to offset such damage, but its reputation and that of the airframe both clearly suffered, with many analysts reporting declining customer confidence. In February 2020, for instance, Boeing's internal research was indicating that 40 percent of travelers would be unwilling to fly on the MAX when it reentered service (Gelles 2019). "Our passengers are nervous," one anonymous airline executive told a journalist, adding that "the expectation is that we will not resume flying the MAX until the industry is satisfied with the fix. But that's going to take time, and that will be damaging for us in the long run as a Boeing 737 MAX customer trying to operate as a competitive airline" (Macheras 2019). Airlines could not ignore such concerns, and by 2020 the 737 program, which once accounted for about 40 percent of the company's earnings, was in serious peril (Topham 2020; Gelles 2019; Isidore 2020b, 2019; Macheras 2019; Mihm 2019; Robison and Newkirk 2019; Nicas and Creswell 2019). There is also good evidence that the crisis significantly affected orders for the next jetliner that Boeing planned to build: the airplane's naming and introduction being delayed while Airbus's rival for the same market segment—the A321-neo XLR—enjoyed strong sales (Isidore 2020b).

Boeing is better positioned to weather this storm today than it would have been a few decades ago, for reasons I will explore, but the accidents will undoubtedly cost the company and its shareholders dearly. In this, therefore, the MAX crisis reaffirms the previous chapter's argument that it is never economically rational for airframers to risk accidents caused by design weaknesses. In supporting that argument, however, the crisis also appears to undermine the subsequent argument: that economic rationality incentivizes airframers to prioritize reliability of their own accord. It is to this tension that we now turn.

IRRATIONAL BEHAVIORS

It is difficult to deny that the MAX disasters are rooted in real failings and compromises by Boeing. Not all accidents are avoidable, as we have seen, and postaccident reportage reflexively paints routine technical practices as wrongdoing. But even so, the accidents do strongly suggest meaningful organizational shortcomings. Boeing's decision not to link the MCAS system to redundant AOA indicators seems to violate core principles of jetliner architecture, for example, and its disavowal of the first accident does not evince a strong commitment to recursive practice. The impression of organizational failings is bolstered further by a slew of other significant safety issues with the MAX that have come to light since 2019, ranging from foreign object debris in fuel tanks (Isidore 2020a) to previously unreported concerns about insufficient wiring separation (Duffy 2020; Tkacik 2019). There is also good evidence, from internal communications and other sources, that most of these issues were viewed as compromises by many of Boeing's engineers at the time, and that they reflected explicit managerial decisions to prioritize profit over reliability (Tkacik 2019; Nicas et al. 2019; Topham 2020; Useem 2019). (The decision to forgo redundant AOA indicators, for example, was reportedly driven by concerns about triggering costly regulatory scrutiny. By the account of one Boeing engineer, the problem lay with the extra management systems that redundant indicators would have required to manage disagreements, which would have required regulatory attention [Tkacik 2019]).

So what should we make of the claim that airframers are driven by economic incentives to prioritize long-term reliability over short-term profits?

In making sense of Boeing's lapses with the MAX, it is important to understand that, even if it is economically rational for airframers to subordinate short-term profits to long-term reliability, it is not necessarily the case that airframers always and infallibly act rationally. Organizations might not be straightforwardly prone to the kinds of cognitive biases that interest behavioral economists (see Harrington and Downer 2019), but they are not immune to irrationality. All are complex, heterogeneous, and imperfect entities, which, as sociologists have long understood, are eminently capable of acting against their own best interests and incentives (e.g., Cohen et al. 1972). Airframers are no exception in this regard, as the story of McDonnell Douglas testifies. Organizational cultures must constantly be reaffirmed and reinvented as their personnel change, and scholars routinely find that even the most valued safety practices tend to "drift" or "migrate" over time unless they are carefully guarded (e.g., Snook 2000; Rasmussen 1997). Economic incentives have driven the reliability of modern jetliners as much by culling errant airframers that failed to prioritize it as they have by keeping such airframers in line (and that the former mechanism played an important role in the latter).

Insofar as the MAX's flaws reflect design compromises made for near-term economic gain, therefore, then it is very possible that Boeing—so successful for so long—simply lost sight of its incentives. In fact, many long-time observers of the aerospace industry maintain that this is exactly what happened. Tkacik (2019) and Useem (2019), for example, both recount a major culture shift within Boeing in the years preceding the accidents, wherein a proud tradition of prioritizing engineering became subsumed by a culture of corporate managerialism, which prioritized supply-chain efficiency and treated jetliners as a typical commodity. By most accounts, this culture shift began with the company's 1997 takeover of McDonnell Douglas;[2] was exemplified by its 2001 decision to relocate its corporate headquarters from Seattle to Chicago, 1,700 miles away from where it builds its jetliners; and was fiercely resisted by its engineers, who even went on strike to protest the changes (Tkacik 2019; Useem 2019).

In grappling with Boeing's relationship to incentives in this matter, it is also worth noting that the relationship between economics and reliability in civil aviation has itself evolved. It is important not to overstate this. To the extent that the company can be said to have subordinated reliability to short-term profit, then it is almost certainly still true to say that it acted against its own economic interests (witness the outsized costs of the MAX debacle, outlined previously). But the balance of that calculation has undoubtedly shifted in recent years.

The principal cause of this shift is the changing nature of competition. Chapter 10 argued that one of the conditions that has long underpinned the extreme reliability of jetliners was a competitive market that makes airframers extremely vulnerable to reputational damage. Yet the exposure of airframers to market competition has diminished significantly in recent decades. There are two basic reasons for this. One is that decades of consolidation and competition have winnowed the industry to the point where there are now far fewer competitive airframers than there were thirty or forty years ago. Really, there are just two in the wide-body market—Airbus and Boeing—both of which are protected from bankruptcy by the fact that they are increasingly seen as vital regional assets. The other is that maintenance and training considerations have increasingly locked airlines into exclusive relationships with one of those airframers (Chokshi 2020). In combination, these trends have meaningfully reduced the market's tendency to severely punish airframers for unreliability. Insofar as Boeing is weathering the MAX crisis, for instance, it is largely because most of its customers have strong incentives to stay with the company for reasons pertaining to maintenance and training arrangements, and because the market understands that the US federal government is unlikely to allow Boeing to go bankrupt.

In this respect, therefore, the MAX accidents might best be understood as the exception that tests the rule. They still testify to the fact that airframers are economically incentivized to prioritize reliability over short-term profits, even if they also remind us of the imperfect sway that those incentives exert on airframers' decision-making.

The industry's long-term economic incentives aren't infallibly determinative of its safety choices, therefore, but they still usually prevail. And should we be tempted to doubt this, the industry's relationship with crash survivability reaffirms the point.

11.3   CRASH SURVIVABILITY

LOST SURVIVORS

In her 2014 farewell speech at the National Press Club, Deborah Hersman, the departing chair of the NTSB, lamented that a "great disappointment" of her ten-year tenure at the agency was that child-safety seats were yet to be required on commercial jetliners (Jansen 2014). The NTSB investigates US aviation accidents to determine their causes, but when it comes to remedial action, the agency only has the power to make recommendations. Under Hersman, it had repeatedly appealed the FAA to mandate that airlines carry child-safety seats, with its experts believing that the measure would save lives. It was not alone in this view. Demands for such seats had been gathering steam for decades, drawing in a wide range of safety advocates. "Every single thing on [an] airplane, down to the coffee pots, are required to be properly restrained except children under the age of two," the president of the Association of Flight Attendants, Patricia Friend, told the New York Times in 2010, adding, "It's just physically impossible, no matter how much a parent loves that child, given deceleration forces of an aircraft in a crash, to hold onto that child" (Higgins 2010). By broad assent, however, the regulator had repeatedly stalled on the issue.

The frustration of safety advocates over child-safety seats is interesting because it is illustrative of a much broader set of tensions around crash survivability (sometimes referred to as "crashworthiness" or simply "survivability") in civil aviation. There is an old adage, usually credited to the American baseball player "Satchel" Paige, that "airplanes may kill you, but they ain't likely to hurt you." The line is intuitive but misinformed. Jetliner crashes involve attempted landings or aborted takeoffs far more often than catastrophic explosions or meteoric plummets, and, as a result, most have injuries and survivors. Contrary to Paige's aphorism, in fact, airplanes are considerably more likely to hurt you than to kill you. Of 53,000 people involved in accidents involving US air carriers between 1983 and 2000, 51,207 of them—96.6 percent—survived (Fiorino 2009).

Of accidents that do produce casualties, moreover, studies have found that upward of 80 percent are potentially "survivable," in the sense that "the crash impact does not exceed human tolerances" (NAS 1980, 9; see also Fiorino 2009; Shanahan 2004; FAA 2016; ETSC 1996). In these accidents, the majority of the fatalities often result from potentially mitigable hazards: evacuation delays, unrestrained impacts, smoke inhalation, fires, cabin missiles, and other incidents. In principle at least, each of these fatalities represents a life that might have been saved if jetliners been designed differently.

It is in this counterfactual pool of "lost survivors"—individuals who might have lived but did not—that critics looking for shortcomings or scandals in civil airframe design and regulation have found their most tractable arguments. As we have seen, the aviation industry will go to impressive lengths to reduce the likelihood of a catastrophic accident even marginally, and yet it often seems far less proactive in its efforts to mitigate the human costs of accidents. Relative to reliability at least, survivability does not appear to be a priority for modern civil aviation.

SAFETY SECOND

As with the claim that Boeing cut corners on the MAX, the claim that civil airframers often fail to prioritize survivability over profit is difficult to deny. FAA-funded studies have concluded that airframers are often unwilling spend more than the minimum required when it comes to survivability measures, even if very modest increases of expenditure promise substantial safety benefits (e.g., Cannon and Zimmermann 1985, 60).

It is true that modern jetliners are, to some extent, designed for survivability. Cabin interiors, for example, are configured to allow passengers to evacuate inside ninety seconds with minimum illumination and half the usable exits blocked. Fuselage doors are designed to open in conditions of extreme icing. Furnishings are designed to be fire resistant. Emergency slides double as life rafts. Seats are equipped with flotation devices and quick-release lap belts and are rigged to withstand strong dynamic forces. Most of these features came relatively late to jetliners, however, and often only after protracted struggles that pitted safety advocates like the NTSB against airframers, operators, and, in many cases, the FAA itself (Anderson 2011; Saba 1983; Bruce and Draper 1970; Nader and Smith 1994; Weir 2000; Perrow 1999).

Examples of these struggles around survivability abound. Perrow (1999, 163–164), for instance, describes a long-running fight to mandate that cabin address systems be equipped with backup power so that the crew can still instruct passengers if the electrical system is damaged. By his account, a series of fatal accidents involving failed address systems and delayed evacuations led to a decade-long lobbying effort by the NTSB to have the FAA require retrofits, each estimated to cost between $500 and $5,000 per aircraft. He tells a similar story about the structural integrity of cabin furnishings, arguing that regulators and manufacturers rebuffed clamorous concerns about outdated and inadequate impact standards for years, while accidents repeatedly tore seats from fuselages (Perrow 1999, 164–165; see also Saba 1983).

Rules governing the flammability of cabin furnishings became a particularly chronic source of contention, with safety advocates pitted against regulators and manufacturers. Three out of four people who died in "survivable" crashes in the years leading up to 2004 were killed by fire, smoke, or toxic fumes (Nader and Smith 1994). A major factor in these deaths was that cabin interiors were full of synthetic materials that, when heated, produced deadly and disorienting gases, such as phosgene and hydrogen cyanide. Alternative materials with reduced fire toxicity were available, but they were slightly heavier and more expensive, so manufacturers resisted efforts to mandate their use (Weir 2000; Perrow 1999). The FAA began studying aviation combustion toxicology in 1972, on the recommendation of the NTSB, and quickly came to the conclusion that cabin materials were producing toxic fumes.[3] However, it wasn't until 2008—a full thirty-six years later—that the regulator issued its first mandatory guidance on passenger cabin smoke protection (FAA 2008b).

These past fights over what became modern survivability standards are echoed today in numerous ongoing struggles. The contest over child-safety seat is a good example, but there are many other measures that safety advocates would like to see implemented. Aft-facing seats are among them. Decades of research has found that reversing the orientation of passenger seats would significantly reduce impact fatalities (FAA 2016; ETSC 1996; Snyder 1982; Cannon and Zimmermann 1985). (This is why military transport vehicles have seats in this configuration.) There are also calls for increasing the seat pitch and equipping cabins with passenger airbags and three-point seatbelts, all of which, it is argued, would reduce impact fatalities (Klesius 2009; Smith 2013). Fires, too, remain a prominent concern. Some safety advocates have lobbied for rules requiring that highly flammable oxygen canisters to be stored below cabin floors, where they would be better protected (Weir 2000, 85). Others have lobbied for jetliners to be equipped with smoke hoods (which have been compulsory on the FAA's own aircraft since 1967, and on most corporate jets and US military transport aircraft since the mid-1990s) (Weir 2000; ETSC 1996); and, to a lesser extent, for cabins to be fitted with sprinkler systems. There are even experts pushing for the use of low-flammability fuels, such as anti-misting kerosene or high-flashpoint kerosene (JP5/AVCAT), which are designed to limit fires and explosions in the event of an accident (Collins 1988; Collins and Pinch 1998; Weir 2000; ETSC 1996). All these measures would have saved lives in the past, but none looks likely to be implemented in the near future.

REGULATION AND FINITISM

Much as with the 737-MAX, the industry's relationship to crash survivability appears to exemplify the claim that regulators are unable to police airframes effectively. In directly pitting the financial interests of airframers against the safety interests of the flying public, survivability measures should be an ideal candidate for regulatory intervention. It is notable, therefore, that in this context, regulators frequently find themselves at odds with safety advocates.

The frequent tensions between the NTSB and the FAA on matters pertaining to survivability speak revealingly to how the inherent ambiguities of engineering knowledge shape the regulator's relationship with the industry it oversees. Examined closely, these tensions do not suggest that the regulator is *unwilling* to defy airframers, so much as that it is often *unable* to do so. It may have taken the FAA decades to implement new fire toxicity regulations, for example, but over that period, it proposed numerous such rules, only to be forced to withdraw them when the industry contested its findings, pointing to uncertainties in the FAA's tests and demanding further studies (Smith 1981; Weir 2000; Perrow 1999).[4]

The uncertainties that airframers invoked to undermine the FAA's proposals were never indefensible, but this is the point. As we have seen, technological investigations—tests, models—always harbor defensible uncertainties, and determining the facts behind a technical question always requires interpretive work: a willingness to accommodate unanswered questions and overrule dissenting voices. Whenever airframers oppose an FAA technical ruling, therefore, they can always point to uncertainty to call for delay and further study. (This is a common tactic of powerful actors eager to maintain the status quo in a wide range of domains [see, e.g., Proctor 1991; Oreskes and Conway 2010].) And the regulator—beholden to a misleading but deeply institutionalized notion of technological positivism, which promises definitive answers to technological questions—is poorly able to deny such calls.

A parallel dynamic echoes in the stories of many other crash-survivability measures, with airframers frequently contesting the FAA's test results and demanding that regulation be postponed until further research has been done. The introduction of independently powered cabin warnings, for example, was delayed by controversies that lasted through the 1970s and into the 1980s (Perrow 1999, 164). The introduction of new standards for seat load factors came after an even longer period of debate: decades of disputed research that, by the 1980s, had the NTSB publicly questioning whether the FAA would ever accept the representativeness of its own test data (Perrow 1999, 165). This dynamic is even more evident in the stories of the many putative survivability measures that died in the laboratory, mired in perennially contested experiments. Studies of cabin sprinkler systems, for example, ran into complex questions about the dangers of steam (Weir 2000, 90). Studies of smoke hoods devolved into protracted debates about the validity of cabin evacuation tests (ETSC 1996; Weir 2000; Flight Safety Foundation 1994). Studies of aft-facing seating became bogged down in questions about impact dangers from underseat luggage (Cannon and Zimmermann 1985, 6). And studies of anti-misting kerosene were largely derailed by debates about engine flameouts (ETSC 1996, 26).

Not all of these debates were wholly unwarranted, of course, and it is likely that some of the measures that died on drawing boards did so for good reasons, but collectively these stories paint a compelling picture of obstruction. (To quote one British accident investigator's candid take on survivability regulations: "The whole point of . . . commissioning research is to kick the issue into touch for a couple of years. Then the research comes back and they say the costs outweigh the benefits. By then, everybody has moved on and forgotten about the . . . original crash in the first place. It is a deliberate tactic. And it works." [Weir 2000, 98]). After all, the FAA has been able to move relatively quickly in matters pertaining to reliability. New bird-strike regulations, for instance, are routinely passed and revised despite their many unanswered questions. In principle, the industry could exploit the ambiguities of technological practice to delay or rebuff such measures—miring them in controversy, as it does survivability measures—but in practice this rarely happens. The industry treats survivability differently from reliability, in other words. And, as we will see, this distinction speaks convincingly to the primacy of economic incentives in shaping jetliner design decisions.

SURVIVABILITY AND INCENTIVES

The history of airframers resisting survivability measures does not intuitively fit with every argument of the preceding chapters. This is because, while substantiating the claim that regulators are poorly positioned to police safety, it simultaneously seems to challenge the claim that economic incentives lead the industry to prioritize safety of its own accord. Considered carefully, however, the industry's adversarial relationship to survivability actually lends a lot of support to the idea that long-term economic incentives have underpinned its commitment to reliability.

To understand why this is the case, it helps to recall a distinction introduced in chapter 1, between reliability (defined as the "probability of a technology failing catastrophically") and safety (defined as the "probability of someone dying in a catastrophic technological failure"). Let us then add to this survivability: (defined as the "probability of someone surviving a catastrophic technological failure, should they experience one.") Now, as has been discussed, the terms reliability and safety are routinely conflated in civil aviation, but, by these definitions at least, safety encompasses more than just reliability. Jetliners are certainly safer if they crash less frequently—so more reliability means more safety (which is why the terms are often conflated)—but they are also safer if their crashes are less fatal. A more reliable jetliner might be a safer jetliner, in other words, but so is a more survivable jetliner.

While reliability and survivability both contribute to safety, however, the incentive structures around making jetliners unlikely to crash (reliability) differ markedly from those around making crashes less dangerous (survivability). The previous chapter outlined why the aviation industry is incentivized to pursue reliability. It argued that the industry pays a heavy price for accidents, especially those that can be tied directly to design choices or manufacturing practices. The same cannot be said about safety for its own sake, however, because even if catastrophic failures can cost the industry dearly, those costs do not scale proportionately with the number of deaths that arise from those failures.

For illustration of this disparity, consider that—from a purely legal, economic, or managerial perspective—it makes little material difference to the key institutional actors in civil aviation (airlines, airframers, or even regulators) if an accident incurs 80 fatalities or 280. The legal implications vary little in each case, not least because it is difficult to hold airlines or airframers accountable for not adopting survivability measures that regulators have not mandated. (And it is difficult to challenge regulators in the courts for not mandating specific survivability measures [Flight Safety Foundation 1994, 18]). The direct economic costs are also broadly equivalent in each case, liabilities for individual deaths being difficult to establish and rarely very punitive. As discussed previously, the real economic costs that jetliner accidents inflict on the industry primarily arise from the reputational damage that accidents cause, and history suggests that this damage correlates only loosely with the total number of fatalities. (Partly because any "lost survivors"—who might have been saved—are difficult to measure, so they tend to remain invisible.)

The upshot of this is that survivability measures, unlike reliability measures, are rarely cost effective from an airframer's standpoint, because the economic return on making jetliner accidents less dangerous is far less than on making accidents less frequent. [5] To understand this calculus it helps to recognize that survivability measures can be counterintuitively expensive, as the direct costs of developing, purchasing, and installing them are often overshadowed by indirect costs that are more opaque. Almost all such measures add weight, sometimes in nonobvious ways, which, as we have seen, can have considerable cost implications. (Aft-facing seats and three-point restraints do not weigh much themselves, for example. In the event of an impact, however, both would raise their occupants' center of gravity from the waist toward the shoulder, creating a leverage effect that would rip seats from the cabin floor without stronger, and therefore heavier, fixtures and structures [Cannon and Zimmermann 1985]). Other survivability measures impose costs by taking up valuable space. (Increasing seat pitch would lessen head impacts, for instance, but it would also decrease the number of seats that airframes can accommodate.) Airlines also fear that some measures—such as aft-facing seats or smoke hoods—would deter customers by disrupting their expectations and hinting at unwelcome possibilities (Cannon and Zimmermann 1985, 4; Snyder 1982).

Understood in this context, therefore, the civil aviation industry's reluctance to invest in survivability, as well as its resistance to regulations that would require it to do so, are easy to reconcile with the argument that its safety behavior is primarily guided by medium-term economic incentives. Indeed, the fact that its behavior is difficult to fault when viewed through the lens of reliability but looks rife with mundane institutional pathologies when viewed through the lens of survivability strongly supports this conclusion. The industry vigorously pursues reliability, we might say, because more reliable jetliners save it money, and it aggressively resists survivability because more survivable jetliners cost it money. Nobody would suggest that the individuals who work in civil aviation are indifferent to its human tragedies. But, as we have seen, individual sympathies have limited agency when it comes to shaping organizational behavior. Ultimately, profit drives airframe design decisions more than safety, and airframers have more control over those decisions than regulators do.

11.4 POSITIVIST PATHOLOGIES

MOVING FORWARD

The stories of the 737-MAX and the industry's relationship to survivability are valuable in that they add nuance and substance to the argument that it is economic self-interest—more than rules, regulators, or good intentions—that polices the designs of modern jetliners. Both cleanly illustrate regulators' inability to effectively police jetliner designs, for example, and both, in different ways, speak to how economic imperatives act on airframers.

In this latter regard, the MAX is a cautionary tale that warns against simple narratives. It illustrates the pressure that competition exerts on technological decision-making, even when experienced organizations like Boeing are making choices about heavily regulated artifacts like jetliners. It demonstrates how that pressure can shape design deliberations, as well as the dangers that can arise from this dynamic. Ultimately, however, it shows why it is still a false economy for airframers to give in to these short-term incentives, even if their managers sometimes lose sight of this fact.

The industry's relationship to survivability, in turn, illustrates that the industry largely understands its incentives. Counterintuitively, perhaps, the resistance to making accidents more survivable speaks compellingly to the fact that economic incentives play a key role in shaping its design decisions. For how better to reconcile the great costs that airframers will shoulder in their efforts to make accidents less likely, on one hand, with their palpable reluctance to invest in making accidents less fatal, on the other, than by understanding that the incentives around these endeavors pull in opposite directions.

Note that examining these stories from this finitist, structuralist perspective offers insights that would not be apparent if they are viewed through a more conventional lens, wherein aviation safety is construed as governed by assessments and policed by regulators. Understood in this way, for example, the MAX crashes raise questions about the changing landscape of aviation economics. It suggests how decreasing competition in the industry, together with airlines' close relationships with specific airframers and Boeing's emergent status as "too important to fail," might all have underappreciated implications for future passenger safety (because they change the economic incentives around reliability). Survivability debates are similarly reconfigured. Seen from this perspective, for instance, laments about child-safety seats might be read less as a regulatory failing than as a reflection, again, of the incentive structures that shape the industry.

These insights, in turn, speak to ways in which the conventional, positivist understanding of aviation safety might itself be a source of danger. As we have seen in this discussion, for instance, the understanding that airframe designs are governed by rules and assessments hides the importance of structural incentives from the policymakers who police those incentives. Or take the industry's problematic relationship to survivability. One of the reasons why regulators and advocates have repeatedly struggled to force the industry's hand in this area is the pervasive but misleading idea that engineering tests should yield objective, definitive, and unambiguous results. Absent this positivist promise of certainty, endless appeals for "further study" to dispel lingering uncertainties would be less effective as a means of delaying or derailing proposed regulatory actions.

Civil aviation's hesitance over survivability has been broadly tolerable over the years because the industry's incentivized commitment to reliability has kept the absolute number of accidents so low: the rarity of plane crashes making the issue of surviving them less pressing. The fact that policymakers might be underappreciating the importance of competition is potentially more consequential, however. And, as we have seen, jetliners have some unique properties relative to other catastrophic technologies.

With that in mind, therefore, let us consider what other pathologies might arise from positivist misconceptions of engineering practice in civil aviation and beyond.

IV   RECKONING WITH CATASTROPHIC TECHNOLOGIES

Wherein it is argued:

- That the misleading portrayal of jetliner reliability management—as fundamentally grounded in engineering tests and models—can be functional, but it also comes with underappreciated costs **(chapter 12)**
- That the practices that experts use to make jetliners ultrareliable depend on resources that are not available to other technological domains, and the most consequential cost of misportraying jetliner reliability management is that it hides this difference **(chapter 13)**
- That reactors exemplify the difference between jetliners and other catastrophic technologies **(chapter 14)**

IV. RECKONING WITH
CATASTROPHIC TECHNOLOGIES

12 BURDENS OF PROOF: THE HIDDEN COSTS OF POSITIVISM

In theory, there is no difference between theory and practice; in practice, there is.
—Walter J. Savitch

For a successful technology, reality must take precedence over public relations, for nature cannot be fooled.
—Richard Feynman

12.1 A THEATER OF OBJECTIVITY

PRETENSE AND HUMBUG

In his memorable introduction to the principles of structural engineering, J. E. Gordon (2018 [1991]) speaks of a divide that characterized the discipline in the early- to mid-nineteenth century. During this period, he explains, structural engineering on the European continent was dominated by a theoretical and quantitative "French" tradition, while a more pragmatic and craft-based "British" tradition ruled on the other side of the English Channel. He goes on to note that the relative standings of these two traditions seem incongruous today. The greatest and most prestigious structures of the age were being built by British engineers like Thomas Telford and Isambard Kingdom Brunel, and yet it was the continental Europeans who pioneered many of the mathematical theories and models that have come to define the discipline. From a modern perspective, Gordon argues, the most lauded engineers of the nineteenth century look almost primitive in their disdain for abstract theory and quantitative analysis (Gordon 2018 [1991], loc. 794–800).

Gordon has an interesting interpretation of this incongruity. He doesn't see it as evidence that French engineers were undervalued by their contemporaries. He sees it, rather, as evidence that British engineers understood something fundamental about their discipline that most modern observers have come to forget. He writes:

> We must be clear that what Telford and his colleagues were objecting to was not a numerate approach as such—they were at least as anxious as anybody else to know what forces were acting on their materials—but rather the means of arriving at these figures. They felt that [French] theoreticians were too often blinded by the elegance of their methods to the neglect of their assumptions, so that they produced the right answer to the wrong sum. In other words, they feared that the arrogance of mathematicians might be more dangerous than the arrogance of pragmatists, who, after all, were more likely to have been chastened by practical experience. (Gordon 2018 [1991], loc. 800–807)

He tells this story as a way of contextualizing and qualifying the formula-rich ("French") theory he is about to explain. It introduces a cautionary motif that runs throughout his book, about structural engineering not being wholly reducible to mathematical tools and formal rules. Engineering knowledge still has vital qualitative dimensions, he wants to say, but the emphasis the discipline places on austere calculations and formal models can seduce us into believing otherwise.

Gordon is far from being the first writer to make this point. Indeed, it is a familiar theme of texts written for lay audiences by academic engineers and observers of engineering, especially when those texts reflect on the nature of the discipline itself (e.g., Petroski 1992a; Blockley 2012). So it is, for instance, that Wynne (1988, 150–151) speaks of a "huge contradiction" at the heart of modern technology governance, wherein a "neat and tidy public image" of engineering practice belies a much "messier" reality. Vincenti (1990), notably, makes a similar argument as Gordon, but in relation to aeronautical engineering, again illustrating it with reference to historical wisdom that gradually came to be forgotten. To this end, he quotes a British engineer speaking to the Royal Aeronautical Society in 1922: "Aeroplanes are not designed by science but by art, in spite of some pretence and humbug to the contrary. I do not mean to suggest that engineering can do without science, on the contrary, it stands on scientific foundations, but there is a big gap between scientific research and the engineering product which has to be bridged by the art of the engineer" (Vincenti 1990, 4).

As Vincenti goes on to argue—and as preceding chapters of this book have illustrated and explored at length—it would not be unreasonable to make the same observation about jetliner design a century later. The process of designing and evaluating ultrareliable airplanes undoubtedly draws on deep foundations of theory, measurement, and calculation (manifest in tests and models), but those foundations cannot sufficiently account for the extraordinary failure performance that jetliners achieve in service. As we have seen, that performance is only really intelligible when also understood in relation to various practical resources and qualitative arts: a deep well of examined experience (thousands of heavily interrogated failures); the tacit expertise required to make complex judgments about relevance; an economically incentivized institutional willingness to consistently prioritize reliability over costs in those judgments, and more. This is why the key to unraveling the "aviation paradox," as I have called it, lay in recognizing that experts have not perfected the accuracy of their tests and models to a level that finitists would find implausible, but circumvent those tests and models by leveraging real-world service experience. They don't "design," "calculate," or "measure" ultrahigh reliability so much as they "whittle," "infer," and "consecrate" it.

As we have also seen, however, jetliner reliability management is not portrayed as functioning in this way. Instead, it is couched in a language of rule-governed objectivity—quantified requirements and calculative practices—that obscures its many subjectivities, indeterminacies, and practicalities, and presents instead an idealized facade of objective rules and deterministic certainty.[1] In this positivist vision of aviation safety, tests and models are capable of revealing the verifiable *truth* of a system's failure behavior, and uncertainty is an aberrant and resolvable condition. By this view, outside experts can *police* the reliability of new jetliners, or any other complex system, to ultrahigh levels, via procedural, rule-governed measurement and analysis. And the public has no more reason to doubt their conclusions than to doubt expert assessments of any other engineering variable.

This is all to say that there is still a lot of "pretense and humbug" in the discourse around jetliner reliability. A tension exists between the way that ultrahigh reliability engineering is commonly understood and the manner in which it actually operates. Jetliners are not safe for the reasons that we believe, and their regulators do not, and cannot, serve the function that we imagine they do. And this tension, in turn, gives rise to a wide range of downstream contradictions and misapprehensions about the function of rules, the abilities of oversight bodies, and the real conditions that foster reliability.

These contradictions and misapprehensions have unanticipated and potentially harmful consequences, which the balance of this chapter will explore. Before that, however, it is worth contextualizing the tension between practice and portrayal a little more.

12.2 PERFORMING OBJECTIVITY

"WHITE-BOXING"

Engineers are far from unusual in overemphasizing the objectivity of their work in public contexts. As a range of literature testifies, it is common for experts to invoke unrealistic abstractions and stylized practices to foster misleading impressions of objectivity and certainty (Lampland 2010; Van Maanen and Pentland 1994; Miller 2003b). This phenomenon is routinely observed in regulatory contexts, where public accountability is important. As Power (2003, 6) puts it, "Regulatory projects are always in some sense visionary . . . possibilities and aspirations for control and order get projected via discussion documents, codes, guidance manuals and the law, often in an ideal form abstracted from the messy realities of implementation."

Nowhere is this more the case than in contexts where scientists and engineers, specifically, explicitly engage with matters of public concern. In these circumstances, it is entirely conventional for technoscientific experts to construct performances of objectivity for lay audiences, hiding their uncertainties and disagreements backstage so as to better present a positivist caricature of their work (e.g., Hilgartner 2000; Rip 1985; Jasanoff 1990, 2003; Wynne 1988, 1989; Collins and Pinch 1998). "[I]f one looks carefully into reassuring public rhetorics of [engineering], one may find hints at a less rule-governed existence for technologies," Wynne (1988, 151) writes, "but, for public consumption, these are highly attenuated and cryptic, if they are expressed at all."

Hence the contradictions outlined here.

Scholars usually understand these stylized performances of objectivity in relation to a modern cultural and legislative milieu that demands formal accountability and idealizes quantitative rules. The emergence of this milieu is explored in an extensive body of scholarship that defies easy summary (e.g., Hacking 1990; Porter 1995; Lampland 2010; Desrosieres 1998;

Verran 2012; Power 1997).[2] Most accounts agree, however, that it has been largely driven by organizational efforts to build legitimacy and authority. Uncertainties invite dissent, and qualitative judgments imply arbitrariness or bias, especially in technoscientific contexts where formal methods have long been idealized. "Non-engineers often seem to stand in awe of the practitioners of applied science and engineering," as Turner and Pidgeon (1997, 14) put it, "regarding them as inhabitants of a world where rationality reigns supreme, where alternative courses of action can be measured and rigorously compared and where science informs every decision."

So it is that masking uncertainties and judgments behind a performance of rule-governed objectivity plays to societal expectations regarding expertise and authority. In doing so, it bolsters the credibility of expert knowledge claims, and thus the power of the organizations that wield them (March and Simon 1958, 165; Espeland and Stevens 2008). "Following rules may or may not be a good strategy for seeking truth," Porter (1995, 4) observes, "but it is a poor rhetorician who dwells on the difference. . . . Better to speak grandly of a rigorous method, enforced by disciplinary peers, canceling the biases of the knower and leading ineluctably to valid conclusions." It is even better, we might say, if those rules are expressed quantitatively, given that numbers—terse, precise, contextless abstractions—are uniquely able to project clarity while simultaneously masking the circumstances of their creation (Espeland and Stevens 2008; Porter 1995).

Note that the argument here is not necessarily that engineers, or experts of any kind, are consciously hiding the unruliness of their work. It is more that a degree of performance has become encoded in the traditions and expectations that frame their stylized discourse. And these traditions and expectations have grown out of longstanding public pressures on organizations to speak authoritatively.

Wynne (1988) neatly captures the essence of this relationship as it pertains to technology oversight in particular, referring to performances of rule-governed objectivity in this context as "white-boxing." The term is intended to convey the irony of a process wherein the inner workings of a system are occluded from public view—as in "black-boxing"—but via processes that are purportedly designed to make those workings visible and accountable.[3] (It is a useful label, as it highlights both the nature and the purpose of the misrepresentation, and I will borrow it for the discussion that follows.)

NECESSARY FICTIONS?

It is worth recognizing that observers sometimes construe the white-boxing of technology, and performances of objectivity in technoscientific work more broadly, as a relatively benign and functional conceit. Ezrahi (2012), for instance, argues that all democratic societies adopt "necessary fictions" in managing their relationship to technical systems.

Defenders of this view usually argue, in different ways, that white-boxing technoscientific claims helps imbue those claims with the authority they deserve. The logic of this is relatively intuitive. As we saw in chapter 2, for example, the indeterminacies of real engineering practices rarely detract from the usefulness and efficacy of expert engineering assertions. When engineers say a bridge will collapse or an engine will overheat, then we ignore them at our peril. So if misportraying those assertions as grounded in wholly objective and rule-governed practices—in conformity with popular perceptions of, and expectations about, engineering work—helps make them convincing, then it can be difficult to see the harm. It is easy to be skeptical of expert claims that are lubricated by the "oily art" of stage management, as Shapin (1995, 255) puts it, but sometimes even the most creditable claims need lubrication (see also Sismondo 2010, 170–173).

By some accounts, white-boxing can also facilitate the *production* and *application* of technoscientific knowledge, as well as its public reception. The argument here is that experts can use performances of objectivity to create backstage spaces, wherein they can more freely and productively exercise important qualitative judgments without inviting censure for not conforming to idealized expectations about their work (Wynne 1998; Espeland and Stevens 2008; Lampland 2010; Schulman 1993). In this view, the hidden ambiguities and interpretive flexibilities of rules and calculations function as a resource. They create slack in the system that experts can use to navigate the subjectivities of their work behind the scenes, without challenging the conventional mores of modern accountability. (So it is, we might say, that aviation experts more easily manage the debates that permeate bird-strike tests because those debates occur backstage, in esoteric technical discourse, where doubts can be voiced and judgments exercised without inspiring public controversy.)

It is easy to imagine that civil aviation, in particular, might exemplify this phenomenon. After all, jetliners might not be reliable for the reasons we imagine, but, as we have seen, decades of service data show that they are as reliable as experts claim (or have been in recent decades, at least). And even if it is impossible for regulators to police that reliability directly, the same service data show that airframers have not exploited this inaptitude in ways that undermine the reliability of their designs. In this context, therefore, there is a strong argument to be made that if the industry and its regulator were to subvert public expectations by being fully transparent about their work, with all its inherent messiness, then they could rob the industry of public confidence that it genuinely merits, and undermine practices that are functioning well.

At the same time, however, any instance where democratic institutions are misleading publics for their own benefit probably merits some introspection. Roads paved with good intentions proverbially lead to dark places, after all. It is important to note, therefore, that even scholars who maintain that performances of objectivity can be functional usually acknowledge that such performances often come with hidden costs. Civil aviation, for example, is in the ironic position of having extraordinarily effective practices for making jetliners ultrareliable, which it misportrays to make them credible. It is difficult to imagine that this is entirely optimal from a governance perspective. As we saw in chapter 11, for instance, airframers we able to exploit this misportrayal to delay survivability measures, invoking the idea that technical questions should have definitive answers to repeatedly call for further research.

With this in mind, let us consider what other hidden costs might arise from performing objectivity.

12.3   THE PRICE OF PERFORMANCE

PRESENTATION COSTS

Any discussion of the costs of white-boxing technological practices would be incomplete without recognizing the straightforward expenditures—of time, effort, and money—that it invariably incurs. Stylized performances of rule-governed objectivity might be useful, but the processes through which they are enacted can be extraordinarily onerous and expensive, especially in catastrophic technological contexts. As we have seen, it is probably fair to say that the work involved in making the reliability of jetliners accountable is substantial, probably costing as much as the work of making jetliners reliable. Understand also that the costs of performing objectivity and formal accountability should not be measured in hours and dollars alone, but also in terms of the less tangible burdens that they place on factors like efficiency, organizational culture, and morale. Such effects are underexplored and difficult to quantify, but a mounting body of scholarship suggests that audit requirements can significantly impede organizations that create complex systems (e.g., Graeber 2015; McCurdy 1993). Vaughan (1996) and McCurdy (1993), for example, both portray NASA as an institution that became progressively weighed down by accountability-driven bureaucracy, to the point where its core mission was jeopardized and it became, in the words of one employee, "the Post Office and IRS gone into space" (Vaughan 1996, 211).

The costs and pressures of formal accountability can also have complex indirect consequences. The National Academy of Sciences (NAS 1998, 47), for instance, found that the expense of establishing a bureaucratically defensible basis for its technical claims had sometimes led the FAA to "work around" its own rule-making processes, addressing safety concerns through nonmandatory guidance or draft recommendations instead of binding rules that would carry more force. (The NAS found this outcome to be less than optimal.) The same report also argued that the slowness of the FAA's procedural determinations sometimes led airlines to delay introducing new safety features, putting them off until they could be sure that their investments would be compliant with forthcoming requirements. The NAS was not impressed by this either, commenting: "The public is not well served by a process that pressures operators to delay implementation of safety enhancements" (NAS 1998, 47). The 737-MAX crisis discussed in chapter 11 also might be construed in this light. Many accounts of the airplane's shortcomings attribute them, at least in part, to compromises born of Boeing's desire to avoid expensive recertification (Nicas and Creswell 2019; Tkacik 2019). Absent regulatory oversight, in other words, it is plausible that Boeing would have chosen to make the MAX *more* reliable.

It is important to tread lightly here, as assessment practices can undoubtedly be functional and some administrative costs are clearly worth paying. Johnson (2001), for instance, draws on the Apollo program to make a compelling argument about the important role that accounting mechanisms can play in structuring and integrating complex technological projects. Hastily made decisions are rarely optimal, meanwhile, and—given the importance of design stability to achieving reliability—it might actually be functional for bureaucratic requirements to stymie innovation. Nobody would seriously advocate removing all the tests, analyses, or documentation that govern catastrophic technologies.

At the same time, however, and without belaboring the point, it is difficult to deny that many technology assessment practices can be dysfunctional; and there are good reasons to believe that this is especially the case in catastrophic-technological contexts. Previous chapters of this book have argued that it is impossible to police the reliability required of jetliners—and, by extension, all catastrophic technologies—with formal rules, and such extreme reliabilities are better understood as being governed by incentives. Insofar as this argument holds, then it is easy to imagine that the extravagant audit requirements these technologies invariably accrue might not always be functional. (Or rather that those requirements serve secondary functions—such as making properties like reliability bureaucratically visible—more than their ostensible function of evaluating reliability and safety.) It is certainly arguable that airframers could, and probably would, make equally reliable jetliners at significantly reduced cost if they did not simultaneously have to make that reliability accountable to an audience that values numbers and the appearance of objectivity. After all, they understand their airplanes better than regulators ever could, and the market ensures that they pay a steep price for unreliability.

Even if there are virtues to be found in making the reliability of catastrophic technologies accountable, therefore, the benefits of this accountability are easy to overestimate and its costs easy to overlook.

CONTRADICTIONS

Beyond the literal costs of performing objectivity are a wide range of more subtle complications that arise from contradictions that performances of objectivity inevitably create, between the work of achieving technological ends on one hand, and the work of *being seen to achieve* those ends on the other. An extensive social scientific literature dating back to Merton (1936; 1940) testifies to the perverse outcomes that can arise when rules and measures imperfectly represent the subtleties of organizational practice. This is a central theme of the sociology of accounting literature (e.g., Miller and Napier 1993; Munro 2004; Neyland and Woolgar 2002; Strathern 2000; Hopwood and Miller 1994; Verran 2012), but it is also a recurrent finding in studies of engineering (e.g., Sims 1999; Schulman 1993).

In these studies of engineering specifically, problems are often found to emerge because some of the audiences for whom performances of objectivity are staged—legislators, for instance—play an active (if usually indirect) role in managing the technologies and practices being represented. In these circumstances, the performances can give rise to unrealistic expectations that then influence decisions regarding the technologies and practices themselves. (Recall again, for instance, the misleading expectation that the FAA's survivability researchers should resolve *every* uncertainty before it enforces new airframe requirements.) Problems can also emerge because real engineering practices are difficult to separate cleanly from their stylized portrayal, meaning that organizations responsible for managing technologies either have to make engineering compromises to make their performances work, or they have to grapple with accountability dilemmas created by poor performance.

To more fully understand how white-boxing's contradictions can cause problems, it helps to consider three broad ways in which those problems can manifest.

HIDDEN TALENTS One way of envisaging how the contradictions that white-boxing creates can cause problems is to think about what happens when decision-makers lose sight of the subjectivities of technical practice. Consider, for example, the way performances of objectivity shape lay perceptions of qualitative expertise. As discussed, the way we talk about engineering makes it easy to forget that engineers necessarily exercise unquantifiable judgment in their work. This oversight comes at a cost. Many scholars, for instance, have found that the convention of portraying technical knowledge as objectively quantifiable tends to devalue any unquantifiable (or unquantified) expertise, and to delegitimize claims that must appeal to such expertise for justification (Wynne 1988; Sims 1999; Silbey 2009; Schulman 1993; Langewiesche 1998b; Jasanoff and Wynne 1998).

These scholars invariably attribute meaningful harm to this phenomenon. Perhaps most famously, for example, Vaughan (1996, 2004, 2005) argues that the 1986 space shuttle *Challenger* disaster might have been averted if NASA managers had not deemed the misgivings of experienced engineers "too qualitative" to merit serious consideration. "Critical information was lost," she writes, "because engineers were trained that intuition, hunch, and qualitative data were inadmissible—'not real science,' 'an emotional argument,' 'subjective'—and any engineer who argued on that basis lost status" (Vaughan 2004, 332).[4]

In devaluing qualitative expertise, moreover, white-boxing simulta-neously devalues many of the qualities—such as experience, judgment, tacit knowledge, and technical intimacy—on which such expertise relies. It fos-ters an image of technical experts as diligent implementers of rule-governed formulas, not judicious exercisers of meaningful discretion: masters of the checkbox rather than the black box. And this, in turn, can encourage out-siders to overestimate the interchangeability of experienced personnel and to underestimate the ramifications of moving or losing them (see, e.g., LaPorte and Consolini 1991; Hopkins 2009, 2010; MacKenzie and Spinardi 1996). As one former aviation engineer cautioned, "the usual corporate par-adigm that all people are essentially interchangeable is a mistake of huge proportion [in civil aviation]. People are not interchangeable, because they hold the institutional memory. Without continuity, safety is sacrificed."[5]

The devaluing of qualitative expertise can also create problems for ini-tiatives like the FAA's program of delegating technical tasks to Designated Engineering Representatives (DERs), as outlined in chapter 10. The designee system has limitations, as we saw, but it undoubtedly serves a valuable func-tion. Insofar as the FAA is able to effectively engage with design decisions at all, it is because the DER program allows it to access the judgment of insiders. Given that public-facing regulatory discourse downplays the value of such judgment, however, the FAA struggles to articulate the necessity of the pro-gram and repeatedly has to defend it against high-level criticism and investi-gation. Nader and Smith (1994, 169), for example, skeptically observe that the FAA "believes in the honor system for airline compliance." (See also, e.g., NAS 1980; GAO 2004; Schiavo 1997; Mihm 2019; Robison and Newkirk 2019.)

Relatedly, and perhaps most consequentially, white-boxing tends to hide the vulnerability of expert judgment to external influences. This is to say that when audiences fail to appreciate that technological experts are making qualitative choices in their determinations and not simply applying objec-tive rules, then they are more likely to overlook factors that might shape those choices, like conflicts of interest, biases, and the circumstances that encourage them. (In civil aviation specifically, for instance, few observers rec-ognize the relationship outlined in the previous chapter, wherein the indus-try's economic incentives tend to favor reliability over survivability.) This is reflected in a widespread belief—organizationally entrenched in many cata-strophic technological contexts—that safety can be governed independent of wider concerns like economics (Perrow 1999; Hopkins 2009). So it is, for example, that finance managers in catastrophic-technological organizations are often given little mandate to consider the safety implications of their decision-making (Hopkins 2009, 2010). And, conversely, safety managers are rarely given much say in the kinds of financial decisions that shape economic incentives.

PRISONERS TO PERFORMANCE   A different set of tensions and contradictions can arise from the practical, real-world difficulty of reconciling engineering's messy qualitative practices with a performance of formal objectivity. We have seen in this discussion that white-boxing technical practices arguably create slack for experts to exercise important qualitative judgments by hiding them. And this might be true in some circumstances, but that space is inevitably limited by what can be effectively hidden.

To invoke the mores of objectivity for this purpose is to exploit the fact that seemingly rigid rules and calculations are more interpretively flexible than is commonly understood. For example, it utilizes the fact that engineers can discretely exercise discretion in how they choose to apply a rule or quantify a variable. Rules and calculations are not infinitely interpretable, however, especially in circumstances where their observance is closely scrutinized. Just as actors in a literal play must monitor their behavior backstage to avoid disrupting the performance onstage, therefore, so technical experts sometimes must make compromises to maintain the performance of rule-governed calculation.

This is to say that technical experts (and the organizations they represent) can find themselves prisoners of their own performances: the requirements of making their work publicly accountable force them to make choices that they privately believe to be suboptimal. A common manifestation of this problem arises when experts find themselves obliged to chase metrics that they know imperfectly represent the phenomena they purport to measure: what Merton (1936, 1940) called "goal displacement." The following interview excerpt, from a UK Ministry of Defence official responsible for avionics quality assurance, neatly illustrates the nature of the dilemma:

> Bid assessment is becoming very formal, and under regulations for government departments it's worse, to the extent—I mean, if we haven't got a valid excuse for why somebody hasn't got a job, then you can take us to court. In fact, my observation is that bidding is not about selecting the winner, it's about justifying why the losers haven't got it. . . . I want to go to certain people because I know

that they're good at their job, and I have brick walls set up internally to stop that happening.[6]

The constraints that accountability imposes on expert discretion can even manifest in the designs of systems themselves. Shatzberg (1999) and Bugos (1996) both illustrate how different military airframes came to be shaped, sometimes suboptimally, by bureaucratic accountability requirements. The same observation might be made about jetliners. Recall, for example, the civil aviation industry's relationship to redundancy (outlined in chapter 6). Given the essential role redundancy plays in *demonstrating*—as distinct from *achieving*—ultrahigh reliability, it is reasonable to imagine that the FAA's audit requirements might lead airframers to favor it over other, potentially more optimal design approaches. (As we have seen, adding redundant elements is not inevitably the most effective way for engineers to spend a system's limited weight or volume budget to maximize its reliability. A single, double-strength element might confer more reliability than a redundant backup, for instance, even while it made that reliability harder to demonstrate quantitatively.) No airframer would publicly concede that its designs are anything less than optimal, but Boeing's assertion (in chapter 6) that two engines on its 777 were safer than four might plausibly be read as an implicit admission that there was a period where the airplane's four-engine configuration owed more to regulatory compliance than design optimization.

Even when the demands of performing objectivity do not restrict experts' discretion directly, they often do so indirectly. As with the Ministry of Defence official, for example, experts will sometimes act against their own private judgment for fear that their work will be challenged retroactively and deemed noncompliant relative to an unrealistic ideal. This dynamic has long been observed in other contexts. Noting that "following the rules" serves as a useful defense against blame, for example, Rothstein, Huber, and Gaskell (2006, 101) describe a process wherein administrative fictions akin to those involved in white-boxing tend to organically morph over time into meaningful organizational expectations, "colonizing" the practices they were intended to facilitate (see also Luhmann 2005, 13; Rothstein and Downer 2012).

MISPLACED BLAME Concerns about retroactive judgment speak to another dilemma that arises from using performances of objectivity to create backstage spaces: the fact that those performances can collapse, starkly revealing the contradictions they hide and fostering impressions more damaging than those they were designed to assuage. When organizations seek to hide the messiness of real technological practice, they often forget that that messiness has a way of leaking out at inopportune moments. This is because performances of objectivity only go so far. Even the most elaborate attempts to white-box technical work are never inviolable, and—with the possible exception of some military spheres—it usually takes little investigative effort to glimpse the less-rule-governed dimensions of catastrophic technologies. As discussed in chapter 4, for instance, it makes sense to think of civil aviation as having a private discourse, wherein discussions of its more subjective practices are hidden from public scrutiny, but the privacy of that discourse is protected more by recondite bureaucracy and engineering esotericism than by substantive confidentiality measures. Such barriers deter most idle scrutiny, but there are nevertheless occasions where outsiders are motivated to look more closely. The introduction of new systems, for instance, often drives coverage that highlights elements of technical practice—especially disagreements—that might otherwise remain invisible. (Recall the controversies that accompanied the launch of Concorde and the A320 with its fly-by-wire avionics.) By far the most acute drivers of unanticipated scrutiny, however, are accidents. Accidents attract media coverage and demand formal investigations, both of which invariably shine lights on practices and determinations that would usually be private.

Regardless of its motivation, such backstage scrutiny routinely uncovers practices that fail to match the white-boxed public portrayal of those practices, and these contradictions, in turn, frequently lead observers to draw misleading conclusions. Arguably the most favorable, or least unfavorable, of these that observers draw from witnessing such contradictions is to construe them as evidence of organizational errors or shortcomings. Recall again, for instance, the safety concerns expressed by software experts about the A320's avionics. As we saw, these concerns were ultimately misplaced—the A320 has been as reliable as any aircraft in history—but they arose, we might say, because the software community overestimated the objectivity of certification. They understood enough to know that the regulator's formal reliability calculations were implausible but were not familiar enough with broader aviation regulation to recognize that such calculations are *always* insufficient, even in nonsoftware contexts, and were not the true foundation on which expert confidence in the system rested.

More commonly, however, and usually more damagingly, audiences wrongly interpret the contradictions of technical practice as evidence of wrongdoing: rule-breaking, for instance, or some other form of culpable negligence. This is especially common in the context of accident investigations, which feed into long-established disaster narratives that are already predisposed to apportion blame (Jasanoff 1987; Wynne 1988). So it is, as Wynne (1988, 149) observes, that accidents routinely inspire "flips" in public attitudes toward technical organizations, where a misplaced image of competence and control becomes an equally misplaced image of disorder and incompetence. Or, as the National Academy of Sciences has put it: "Following some highly publicized accidents, there is a technically unjustified loss of public confidence" (NAS 1998: 45).

Langewiesche (1998a; 1998b) captures the essence of this dynamic in his account of the 1996 crash of ValuJet Flight 592 in the Florida Everglades. The accident occurred when incorrectly stowed oxygen canisters caught fire in the airplane's cargo hold, creating an inferno that engulfed the main cabin. Investigators and the media largely blamed this on airline ground personnel, who, they argued, failed to follow correct procedures and falsified paperwork when loading the canisters (NTSB 1996). By Langewiesche's account, however, the underlying cause is more reasonably attributed to ambiguities in the procedures themselves, and the decision to place all the blame on the ground personnel speaks to the aforementioned contradictions between engineering practices and their public portrayal. He writes:

> The falsification [that ValuJet's ground personnel] committed was part of a larger deception—the creation of an entire pretend reality that includes unworkable chains of command, unlearnable training programs, unreadable manuals, and the fiction of regulations, checks, and controls. Such pretend realities extend even into the most self-consciously progressive large organizations, . . . The systems work in principle, and usually in practice as well, but the two may have little to do with each other. Paperwork floats free of the ground and obscures the murky workplaces where, in the confusion of real life, . . . accidents are born. (Langewiesche 1998b, 97)

The same dynamic can affect regulators themselves, with the authority born of couching their work in the language of quantitative certainty sometimes fostering an image of incompetence when their assertions fall short. As previously discussed, for instance, regulatory failure is a prominent theme of coverage around the 737-MAX (e.g., Mihm 2019; Robison and Newkirk 2019), and of the official investigation into Deepwater Horizon (NCBP 2011, 56–57).

Again, it is important to tread carefully here. Attributions of negligent behavior are not always unwarranted. There have undoubtedly been many accidents caused by organizations and individuals who should have been more diligent in designing, building, regulating, and operating catastrophic technologies. These actors are not beyond reproach, therefore, and should in no way be immune to charges of culpable shortcomings. Nevertheless, it is also true that charges of wrongdoing sometimes say less about real failings—organizational or individual—than about the way that technical work is portrayed and perceived. As we have seen, civil aviation's remarkable successes, as well as its very occasional failures, are largely born outside the formal rules, and, contrary to the implicit promises of white-boxing, some accidents are not easily attributed to wrongdoing or negligence (especially those that can be described as "rational" or "normal").

When blame is misplaced, moreover, it can be harmful in ways that go beyond injustice, not least because organizations often respond to charges of wrongdoing in ways that are themselves perverted by the same misconstruals of technical practice. Perhaps most straightforwardly, for instance, it is common for experienced personnel to lose their jobs in the wake of accidents—the symbolic sacrifice of high-placed careers having become a ritual of modern technological disaster investigations—and, when unjustified, this has costs that organizations underestimate due to white-boxing's elision of qualitative judgment. It is also common for organizations to respond to misplaced blame in ways that exacerbate white-boxing's contradictions and tensions (which, ironically, foster the misplaced blame). Blind to the less rule-governed dimensions of technical practice, for example, organizations frequently respond to perceived failings by expanding their formal oversight requirements, deepening their performances of objectivity by drafting even stricter, less practicable rules and requirements. (The Deepwater Horizon report, for example, advocates for the creation of "strict [safety] policies requiring rigorous analysis and proof" [NCBP 2011, 126].) It is also common for organizations to reform their practices to make them better match their idealized portrayal of those practices, further constraining experts' discretion by demanding that unrealistic rules and suboptimal requirements are actually observed (NAS 1998; Francis 1993; Sismondo 2010, 143; Rothstein et al. 2006).

12.4   MITIGATED CONSEQUENCES

NO HARM, NO FOUL?

The significance of the burdens and contradictions outlined here varies, but it is generally muted in the context of civil aviation. The direct costs of making jetliners accountable undoubtedly create inefficiencies and have other complex consequences for the industry, as does the tendency to undervalue subjective judgments (and the experts capable of making them). At the end of the day, however, neither phenomenon has prevented the industry from making jetliners ultrareliable.

Much the same could be said about the tendency to misapportion the blame for accidents and respond to that blame in counterproductive ways. Civil aviation has seen injustices and inefficiencies because of this, and—given its propensity to progressively exacerbate white-boxing's contradictions—this dynamic could lead to greater problems in the future. As with the issues described in this chapter, however, the evidence suggests that it has not yet had a huge effect on the industry's safety (in part, no doubt, because its effects are greatly mitigated by the fact that jetliners crash so infrequently).

Probably most worthy of concern is white-boxing's tendency to hide the role that economic incentives play in underpinning the industry's safety. Insofar as the costs of the 737-MAX's failings are mitigated by declining competition between airframers, for instance, this is a meaningful concern. It will take time for this relationship to become fully apparent, however, and it is probably too early to speak definitively about fundamental changes to the industry's incentives.

Civil aviation's practices might not be transparent, in other words, and they might not be perfect, but this should not eclipse its undeniably impressive service record. By any reasonable definition, jetliners are safe to fly and have been for decades. Jetliners are not like other catastrophic technologies, however, and, as chapter 13 will explain, the real harm that arises from white-boxing their safety is that it conceals this difference.

13 THE MYTH OF MASTERY: ON THE UNDERAPPRECIATED LIMITS OF TECHNOLOGICAL AMBITION

The history of past investigation demonstrates that greater prudence is needed rather than greater skill. Only a madman would propose taking greater risks than the great constructors of earlier times.

—Wilbur Wright

Beware of false knowledge; it is more dangerous than ignorance.

—George Bernard Shaw

13.1 FALSE PROMISES

TECHNOLOGICAL MASTERY

When America's least loquacious man stepped onto the Moon in 1969, most of his spectators appreciated the extraordinary dangers involved in getting him there. NASA had spoken repeatedly about its efforts to ensure Neil Armstrong's safety, but it had also been relatively forthright about the risks he faced. A great many would have been saddened if the *Apollo* 11 astronauts had not returned, but few would have been terribly surprised.

It is doubtful whether the assurances and expectations that accompanied the first Moon landing would have been quite the same if it had occurred today. Publics now expect organizations like NASA to promise more from complex technologies, especially in matters pertaining to safety. Many factors have contributed to this shift in expectations, but it has undoubtedly been facilitated by an increasingly caricatured portrayal of engineering safety assessments as quantitative, objective, and definitive. By promulgating the idea that experts can interrogate technical systems with deductive certainty, the routine white-boxing of complex technologies in public discourse has made possible, and even encouraged, a corresponding belief that experts can make firm promises about the failure behaviors of complex systems. This belief underpins a deep-rooted sense of technological mastery—a widespread and institutionalized conviction that engineers can, and therefore *should*, know the reliability of the machines they build.

On one level, the premises underpinning this sense of mastery are fundamentally misleading—experts cannot perfectly interrogate their machines—but as we saw in chapter 2, the conceit is broadly functional in most contexts. The levels of reliability that experts claim of most technologies are not high enough for the epistemological limits of proof to make a material difference to those claims' usefulness. (It being much more practical, when designing, testing, or modeling systems, to anticipate circumstances that might occur every 10,000 hours, than to anticipate those that might occur every billion hours, especially when those systems are relatively straightforward.) No technology is entirely immune to rational accidents—the hidden uncertainties of technical practice always harbor the potential for rare but unexpected failures—but such failures are too rare to matter in most contexts.[1] So it is that a belief that experts can speak decisively and authoritatively about reliability has long contributed to a practical relationship between engineering and civil society. We might come to regret our technological creations for any number of reasons—the pollution they cause, the disruptions they bring, the costs they incur—but those creations usually function about as reliably as experts promise. And if they don't, it is usually appropriate to impute some level of avoidable error or shortcoming to the various experts responsible for them. Broadly, our intuitions, structures, practices, laws, and assumptions are all well adjusted to navigating most matters related to the probability of technological failure.

It is only in the context of what I have called catastrophic technologies— where the viability of complex systems depends on them being ultrareliable— that the inherent indeterminacies of knowledge threaten to meaningfully undermine the authority of expert claims about reliability. This is because, as we have seen, the fundamental uncertainties of engineering knowledge become uniquely consequential when making assurances about billions of hours of failure-free operation. In these circumstances, even extremely rare accidents can dramatically shape reliability calculations, so even the most trivial uncertainties—born, perhaps, of marginal misjudgments about the equivalence between test conditions and real-world environments—become determinative. Here, therefore, where engineers work in the shadow of disaster and must establish extreme reliabilities in advance of service data, finitist dilemmas like the problem of relevance have a real bearing on the viability of their ambitions and the credibility of their assertions. In this context, if in no other, the idea of engineering mastery over reliability becomes misleading and potentially dangerous. No amount of formal analysis—testing and modeling—can realistically guarantee that a complex system will operate for billions of hours without encountering rational accidents.

UNDERAPPRECIATED

The unique difficulties that arise when complex technologies demand extreme reliabilities are important to recognize because they are systematically underappreciated. Reliability is the sine qua non of catastrophic technologies (almost by definition), but the way we speak about it and govern it in catastrophic-technological domains rarely, if ever, reflects the unique epistemological challenges that it poses.

There are several reasons for this, each of which has been outlined in preceding chapters. First and most fundamental, it is because the distinctiveness of engineering's relationship to catastrophic technologies is counterintuitive and nonobvious. As discussed previously, most of the claims that reputable engineering bodies make about most technologies are manifestly and deservedly credible, including most of those made about technological reliability. When respected experts make assertions about the properties of a system—be it "this engine will burn x amount of fuel per minute" or "this building will not be structurally sound"—it is wise to listen. The claims that they make about ultrahigh reliability in catastrophic technologies are an exception to this rule. These are much less credible than most other engineering claims, as we have seen, but it is far from obvious why this would be the case. Our heuristics about the authority of engineering expertise fail us in this context.

Second, it is because very few catastrophic technologies accrue enough service data for any deficit in their reliability to become self-evident. Any technology that is built and operated in limited numbers—tens or even hundreds of units—would still fail incredibly infrequently, even if its mean-time-to-failure was 100 times, or even 1,000 times, less than we require of catastrophic technologies. Where a dearth of service experience makes extreme reliability impossible to fully master, in other words, it simultaneously hides that lack of mastery from the public. Where catastrophes do still occur in these circumstances, they do so with such rarity that they can invariably be explained away, blamed on specific design weaknesses or organizational shortcomings instead of being understood as evidence of fundamentally intractable problems (Downer 2014; Hilgartner 2007).[2] Ultimately, humans are predisposed to think at human scales (Kahneman 2011), so any system that operates failure free for decades intuitively feels like evidence of ultrahigh reliability, even if its required mean-time-to-failure is north of a million years. (Hence the absence of civilization-ending atomic wars is routinely cited as evidence that deterrence networks have obviously made us safer.)

A third reason why we underappreciate the difficulties of making catastrophic technologies ultrareliable is that the only such technologies that we operate at enough scale for their reliability to be statistically (and intuitively) visible are jetliners, and jetliners, confoundingly, have proved to be as reliable as experts have promised. So the fact that reliability in jetliners is ostensibly governed in the same way as in other catastrophic technologies—rigorously interrogated via tests and models, under the watchful gaze of a dedicated regulator—makes their remarkable (and highly contingent) failure performance look misleadingly generalizable. If jetliners were crashing 100 or 1,000 times more frequently, then it is reasonable to assume that publics and policymakers would more intuitively appreciate the limits of engineering knowledge and the implications of those limits for technologies like reactors and deterrence networks. With this final consideration in mind, therefore, it is worth pausing to recap why it is, exactly, that the reliability of jetliners is so misleading.

FALSE EQUIVALENCE

As we have seen at length, the extreme reliability of jetliners is misleading because it was not achieved in the positivist manner that we are led to believe via formal analysis and rule-governed oversight. Airframers evolved that reliability gradually, by exploiting real-world experience on a massive scale to whittle the uncertainties of an uncommonly stable design paradigm. Regulators honed their assessments in much the same fashion, learning from hard experience how frequently jetliners built around that paradigm failed in real-world operations. In both cases, the process took decades, exacting an enormous cost in accidents and lives. It was practicable only because jetliners were built and operated in far larger numbers, and with far deeper commonalities, than other catastrophic technologies. These commonalities were only sustainable organizationally because civil aviation's operating volume—among other conditions—gave rise to an uncommon incentive structure, wherein perceptions of unreliability were heavily and quickly punished financially. And it was endurable politically only because public risk tolerances evolved with the industry itself, a process that was possible only because jetliners are chronic catastrophic technologies; their extreme reliability demands being driven as much by their volume as by the catastrophic potential of any individual failures.

To be more methodical, the preceding chapters outlined four conditions on which the process outlined depends:

1. *A long legacy of expansive service from which to learn and on which to build*: billions of cumulative operational hours, born of decades of flight by tens of thousands of jetliners with substantial design commonalities, that created a deep well of experience, including a slew of catastrophic failures, which experts could mine extensively for the marginal, esoteric insights required to hone their designs and assessments.

2. *A longstanding commitment to recursive practice*: an institutionalized willingness to invest heavily in the difficult and expensive work of mining the service experience in the manner outlined here, exploring and generalizing from failures and near-misses.

3. *A longstanding commitment to design stability and innovative restraint*: an institutionalized willingness to forgo the promises of radical innovation to aggregate the lessons of the past, manifest as a commitment to a common design paradigm that airframers deviate from rarely, incrementally and extremely carefully.

4. *Structural economic incentives that support these longstanding commitments*: a set of circumstances—such as the existence of competition in the industry, and a tendency of passengers to punish failures in their purchasing decisions—which, in combination with civil aviation's expansive service, tend to quickly punish any systemic unreliability.

These are all *necessary* (albeit, again, not *sufficient*)[3] conditions. Remove even one, and the challenges of ultrahigh reliability become prohibitive. And they all have significant interdependencies. The design stability contributes to the accrual of relevant service experience, for example, and makes the lessons learned from it more useful, for example, while the service experience is crucial to the incentive structure and provides the lessons from which to learn.

It is consequential, therefore, that no other catastrophic technology enjoys the same confluence of conditions. From deterrence infrastructures to drilling platforms, reactors to financial networks, few enjoy even one of these conditions to the same degree as jetliners. (This shouldn't be surprising, given that their interdependencies make each more difficult to achieve without the others.) None are designed with an equivalent commitment to design stability or are operated in equivalent numbers, so none accrue service experience at the same rate as jetliners. This dearth of service experience, in turn, means that the experts who manage and build these technologies have far fewer accidents to mine for the subtle but dangerous misunderstandings that lurk in the epistemology of their designs and assessments. It also means that any insights that those experts do glean from service are often rendered moot by changing designs. And, beyond that, it means that the incentive structures that frame their work are unlikely to consistently favor ultrahigh reliability, with the many costs and sacrifices it demands. (The dearth of service experience allowing reliability shortfalls to remain hidden for long periods.)

Most catastrophic technologies could not enjoy these conditions even if societies were committed to re-creating them. There is no world in which we could build and operate as many reactors or ultradeep drilling platforms as there have been jetliners, and no world in which we could let them fail in the same way. Chapter 1 outlined two types of catastrophic technology, each requiring comparable levels of failure performance, but for slightly different reasons. These were (1) chronic catastrophic technologies, which can tolerably be allowed to fail on rare occasions but require ultrahigh reliability because of the volume at which they operate; and (2) acute catastrophic technologies, which operate at much smaller volumes but require ultrahigh reliabilities because their failures are wholly intolerable. Jetliners are chronic catastrophic technologies: we operate them in large numbers and broadly tolerate very infrequent accidents. Most other catastrophic technologies— deterrence networks, reactors, financial instruments, or drilling platforms— are acute. We build them in much smaller numbers but with comparable reliability demands because they can *never* be allowed to catastrophically fail (as the consequences of such failures could be truly intolerable). There can be no trial-and-error learning with such systems; no "searching for safety," as Wildavsky (1988) once put it.

So it is that civil aviation's technique of ratcheting systems to ultrahigh reliabilities is simply not viable in most other catastrophic-technological domains. In these domains, therefore, experts pursuing extreme reliabilities must *actually* operate in the positivist manner that those in civil aviation only purport to operate, with all the epistemological perils and limitations this implies. This is to say that their designs, and any assessments of their designs, have to be wholly grounded in knowledge gleaned from tests and models. And any shortcomings in those tests and models—born perhaps of imperfect relevance assumptions—are liable to become shortcomings in their designs and assessments, each a potential rational accident. The organizations that structure this work also have to operate against their own economic incentives, which rarely reward the kind of expenditures and sacrifices that extreme reliability requires.

Bluntly put, the reliability achieved in jetliners is unlikely to be achievable elsewhere. Civil aviation's service record demonstrates that modern societies are, in principle, capable of building a complex sociotechnical system that predictably operates for billions of hours between fatal surprises. What it does *not* demonstrate, however, is that this achievement is necessarily transferrable. Jetliners are evidence that experts have mastered extreme reliability in a very specific design paradigm, not that experts have mastered the tools and processes for achieving ultrahigh reliabilities more generally. In this regard at least, most catastrophic technologies are in no way equivalent to jetliners, and the reasons we trust the latter should give us cause to doubt the former.

The most serious harm that arises from white-boxing jetliner reliability is that it hides this difference. Presenting civil aviation's record as grounded in objective, positivist processes hides the informal foundations—conditions and commitments such as design stability, service data, and recursive practice—on which jetliner reliability is built; and masks the price at which that reliability was bought. In doing so, it obscures the uniqueness of civil aviation, and with it the costs and difficulties of replicating its reliability achievements elsewhere. It gives us no reason to doubt the efficacy of tests and models, no reason to look too closely at structural incentives, and no reason to imagine that extreme reliabilities *need* to be incubated over long periods of unreliability. Instead, it suggests that any complex system, if designed, governed, and assessed with equivalent quantitative rigor, might perform as reliably as a modern jetliner (and expert assertions about that system's reliability might be as credible).

The fact that the reliability issues pertaining to all catastrophic technologies are portrayed in the same way in all catastrophic-technological domains, and are ostensibly managed in the same fashion—via rules and standards that call for careful measurements and are enforced by objective regulators—colludes in this illusion. So does the limited service experience that most catastrophic technologies accrue: an apparent absence of accidents (born of sparse service) making it easy to believe (and difficult to disprove) that unrealistic reliabilities have been achieved.

By attributing the reliability of jetliners to structures and practices that look equivalent across domains, the white-boxing of aviation safety invites us to use aviation as a touchstone by which to gauge our mastery of catastrophic technologies. In doing so, it creates a dangerous misapprehension. It is an irony of modernity that the only catastrophic technology with which we have real experience, the jetliner, is highly unrepresentative, and yet it reifies a misleading perception of mastery over catastrophic technologies in general. Meanwhile, that misleading perception of mastery itself helps hide the unrepresentativeness of jetliners.

13.2 THE PERILS OF CERTAINTY

TECHNOLOGIES OF HUBRIS

The widespread misapprehensions outlined in the previous section have made modern societies poor at reckoning with catastrophic technologies. Unlike most other expert engineering assertions, those made about the reliability—and thus the safety and viability—of our most dangerous creations are simply not credible. This is hidden, however, by an idealized understanding of engineering knowledge, deeply entrenched in our institutional logics: an understanding implying that experts speak about extreme reliability with the same authority as they speak about other engineering variables, and that they can achieve such reliabilities in complex systems without gradually whittling them from painful service experience.

This misplaced sense of mastery, with its implausible promise of extreme reliability, is dangerous. There is an old adage that "great things are achieved by those who don't know that failure is inevitable," and while this may be true, the same cannot be said of prudent policymaking (or insightful scholarship) with regard to technologies that cannot be allowed to fail.

The dangers of this promise are accentuated by its reach and invisibility. It permeates discourses and decision-making around catastrophic technologies; explicitly—or, more often, implicitly—underpinning a range of consequential academic, legal, strategic, administrative, and legislative edifices.

In academia, for example, it is heavily implicated in various influential studies, from historians, sociologists, psychologists, and others, that treat expert assertions about the safety of catastrophic technologies as incontrovertible facts, to then be contrasted with (seemingly irrational) public perceptions of risk (e.g., Slovic 2012; Weart 1988; Douglas and Wildavsky 1982; Erikson 1991; Sjöberg 2004; Taebi 2017; Starr 1969). It is similarly implicated in analyses comparing the economic or environmental costs of various energy options, which almost never consider the possibility of reactor failures; in security scholarship that strategizes nuclear weapons deployment with little thought about potential accidents; and in myriad other judgments across a wide range of disciplines.

More materially, and arguably more consequentially, however, this promise is manifest in the decisions that such scholarship informs; the strategic choices societies make about what they build and how they build it.

Consider, for example, recent trends in petroleum extraction. A decade after the loss of Deepwater Horizon, with its extraordinary harms and associated costs—over 200 million gallons spilled and tens of billions of dollars in cleanup costs alone—ultradeepwater drilling platforms have continued to proliferate. By 2017, they were producing 52 percent of all oil in the Gulf of Mexico, up from just 15 percent in the decade leading up to the disaster (Murawski et al. 2020). Oil companies are deploying these platforms in ever-deeper waters—up to twice the depth at which Deepwater Horizon operated—and are drilling ever deeper into the ground, where temperatures and pressures are even more extreme. The backdrop of these trends is an attitude, said to be pervasive within the industry, that evolved quickly from "An accident like Deepwater could never happen," to "An accident like Deepwater could never happen again" (Calma 2020).

Consider also the recent burgeoning of "synthetic credit products": complex financial technologies like the collateralized debt obligations (CDOs) and credit default swaps that failed in 2007–2008, almost bringing the entire global financial system to its knees (Gorton 2012; King 2016). For all the public recrimination that that crisis prompted, there is little evidence that societies have learned to be appropriately wary of the innovations that allowed it to occur. Trade in CDOs—which Warren Buffett presciently referred to as "financial weapons of mass destruction" in a 2002 shareholder letter (Graffeo 2021)—quickly resumed after a postcrisis lull (Boston 2019). Perhaps more ominously still, CDOs have since been eclipsed by another synthetic credit product called "collateralized loan obligations (CLOs)." Similar to CDOs but built on loans made to (often highly troubled) businesses rather than homeowners, CLOs have equal catastrophic potential, if not more (Partnoy 2020). The 2007 market for CDOs is estimated to have been around $640 billion; the 2020 market for CLOs was estimated to be over $870 billion (Partnoy 2020). Should a large number of risky businesses fail at the same time due to an unforeseen common-cause failure—another pandemic, for instance— then this market could collapse, again exposing the banks. Yet there has been minimal preparation for any such catastrophe. Banks deemed "too big to fail" in 2008 have grown even larger in the intervening years. (J. P. Morgan, for example, doubled in size between 2008 and 2018.) And regulations put in place to improve the industry's resilience have started to be rolled back (Li 2018).

Beyond drilling and banking, an even more compelling illustration of the promise of implausible reliability at work is offered by atomic weapons— arguably the original catastrophic technology. The strategic decision-making around such weapons rarely, if ever, grapples meaningfully with the possibility of catastrophic failure; a convention that has persisted through a litany of close calls with disasters that almost defy comprehension (Sagan 1993; Schlosser 2013).

These close calls include a slew of chilling incidents involving individual warheads. In 1961, for example, a technological failure led the US Air Force to accidentally drop two hydrogen bombs on Greensboro, North Carolina. One was only kept from detonating by a small, damaged switch, one of four redundant safety measures, which a recently declassified investigation claims might easily have shorted (Pilkington 2013; Schlosser 2013, 246).[4] A similarly alarming event occurred in 1980, at a Titan II missile silo in Damascus, Arkansas, when a dropped wrench led to an explosion that catapulted a 740-ton silo door into the air, together with a 9-megaton nuclear warhead (Schlosser 2013, 392–398). The warhead—said to have been three times more powerful than every bomb dropped during World War II combined—is again now thought to have been at meaningful risk of detonation (Scholsser 2013, 440; O'Hehir 2016). These are not isolated examples. Historians are uncovering a chilling number of potentially catastrophic malfunctions and mishaps with atomic weapons as the mists of secrecy slowly lift on the Cold War. Schlosser (2013) reports close to 1,200 dangerous events occurring just between the years 1950 and 1968.

Even more alarming, insofar as that is possible, are a slew of near-misses arising from technological failures in the deterrence system itself—failures that might have caused atomic wars rather than isolated explosions. Again, these events are shrouded in a veil of secrecy, such that historians know a lot more about the earlier years of the Cold War about later years. In that period alone, however, they have identified a slew of occasions where false alarms about Soviet missile launches—many caused by technological failures— mobilized the fast-moving machinery of US nuclear retaliation, only for it to be stood down before any commitment became irrevocable. In October 1960, for example, an early-warning radar system in Greenland misinterpreted the Moon as a major incoming Soviet missile strike (Stevens and Mele 2018; UCS 2015). (It created a panic that could have been significantly worse if Soviet premier Nikita Khrushchev hadn't serendipitously been in New York at the time.) In November of the following year, a failed relay station led US Strategic Air Command Headquarters to lose contact with the North American Air Defense Command (NORAD) and multiple early warning radar sites simultaneously, a situation only thought possible in the event of a coordinated attack (UCS 2015). In November 1965, a mass power outage combined with a series of malfunctions in bomb-detecting equipment created a compelling illusion of nuclear attack, prompting another major alert (Philips 1998). In May 1967, radar interference from a solar flare was interpreted as intentional Soviet jamming intended to cover for a nuclear attack, almost leading to a counterstrike (Wall 2016). Jumping forward to November 1979, a computer error at NORAD headquarters led the agency to inform the US national security advisor that the Soviet Union had launched 250 missiles (a figure that it subsequently revised to 2,200) at the US, and to announce that it needed a decision on retaliation in three to seven minutes (Philips 1998; Stevens and Mele 2018).

By general consent, however, the most dangerous such incident known to historians came from the Soviet side. On September 26, 1983, one of the

Soviet Union's early warning systems mistook sunlight reflected off clouds as a US missile launch. The malfunction came at a time of significant tension and distrust between the Soviet Union and the US, and many believe that, had the duty officer at the radar station, Stanislav Petrov (whose name deserves to be remembered), reported the launch as protocol dictated, then a retaliatory launch (and ensuing thermonuclear war) would almost inevitably have followed. Civilization rolled two sixes, and Petrov, who distrusted the technology, chose to defy protocol and wait for further confirmation (Myre 2017; Schlosser 2013, 447–448). As General Lee Butler, a former head of US Strategic Command, put it, "we escaped the Cold War without a nuclear holocaust by some combination of skill, luck and divine intervention—probably the latter in greatest proportion, . . . [b]ecause skill and luck certainly don't account for it" (quoted in Kazel 2015).

This is all to say that we have long been tempting fate, and the age of catastrophic technologies is only in its infancy. As Beck (1992; 1999) and others have observed, decades of near-misses and painful lessons have not meaningfully lessened the confidence with which we confront the risk of technological failure. A host of new catastrophic technologies are emerging without eliciting appropriate levels of concern in this regard. Entranced by the promise of implausible reliability, and implausible certainty about that reliability, our appetite for innovation has outpaced our insight and humility.

The technologies discussed here speak to this dynamic. The fact that sunlight reflecting off clouds could have instigated the end of the world as we know it probably should weigh more heavily on our understanding of progress. But nothing exemplifies and illuminates the issue better and more clearly than the story of nuclear reactors.

By way of a coda, therefore, let us return, finally, to Fukushima.

14 FUKUSHIMA REVISITED: REAPING THE WHIRLWINDS OF CERTAINTY

Eight years involved with the nuclear industry have taught me that when nothing can possibly go wrong and every avenue has been covered, then is the time to buy a house on the next continent.

—Terry Pratchett

14.1 GOVERNING REACTORS

WOEFULLY UNPREPARED

As the sun set on the first day of the Fukushima disaster in 2011, the embattled on-site operators were grappling with mundane supply shortages. Some searched nearby homes for flashlights so they could monitor crucial gauges through the night. The plant did not have its own supply. Others scavenged among the site's flooded detritus for batteries so they could maintain contact with the emergency response center. Their mobile communications units had been equipped with only one hour of reserve power. Nobody had planned for a catastrophe of the scale they were facing. Similar failures of foresight were manifest at every turn. Personal dosimeters were in short supply. On-site dosimeters maxed out at levels far lower than required. The plant's *Severe Accident Manual* lacked appropriate guidance. The designated off-site command post had no air filters or reserves of food, water, or fuel. The region around the plant was even less prepared. Protective medications were scarce, and their distribution was hindered by byzantine rules. Evacuation procedures and official announcements were unrehearsed and poorly considered. There can be no such thing as an orderly nuclear meltdown, but by any reasonable standard, Japan was woefully underprepared for the crisis it was facing (Perrow 2011; Osnos 2011; McNeill and Adelstein 2011; Lochbaum, Lyman, and Stranahan 2014; Kubiak 2011; Onishi and Fackler 2011).

It is tempting to blame Japan for its inadequate preparation, but few if any of its peers were substantially better prepared for such a crisis. In the US, for example, reactor emergency and evacuation planning was (and largely remains) similarly deficient, having been premised almost exclusively on responding to small leaks rather than major meltdowns (Clarke and Perrow 1996; Perrow 2011, 46–47; Kahn 2011). Indeed, the inadequacy of US preparation became evident during the Fukushima crisis, when its tools for assessing radiological threats to its embassy maxed out at fifty miles (Lochbaum et al. 2014, 63).[1] There have been global efforts to improve preparations for major reactor accidents post-Fukushima. Such efforts are limited, however, and it is fair to say that the accident highlighted a blind spot for disaster that still characterizes atomic energy worldwide.

This blind spot is inscribed in the very landscape of atomic energy. Take, for example, the siting of various reactors upwind of major metropolitan areas, such as Indian Point, thirty-five miles north of Manhattan, or Hinkley Point, about the same distance from Bristol in the UK. Or consider that, for economic and political reasons, most nuclear plants are clusters of multiple reactors, meaning that a major failure in one reactor can jeopardize its neighbors, as occurred at Fukushima (Perrow 2007, 136; Osnos 2011, 50). No society would make such choices if it understood Fukushima-scale accidents to be a meaningful possibility. And, having made such choices, no society could adequately prepare for such accidents, even if it resolved to do so. Cities on the scale of Tokyo, New York, or even Bristol are not easily evacuated. Cities of any scale are not easily evacuated. If at first you don't succeed, an old joke goes, then nuclear energy isn't for you.

PRACTICALLY UNTHINKABLE

Seen through the lens of the preceding argument, the global blind spot for reactor catastrophes reflects the positivist promise of implausible reliability,[2] and Fukushima exemplifies the dangers inherent in that promise. Simply stated, states don't plan for such catastrophes because they are confident that such catastrophes will never occur (Rip 1986; Clarke 2005; Fuller 1976; Kahn 2011; Lochbaum et al. 2014; Wellock 2017). The US, for instance, draws a distinction between credible reactor events, such as small radiation leaks, which it deems worthy of public concern, and hypothetical events, such as Fukushima-level meltdowns, which it does not. Japan is no different. This is why Fukushima's Comprehensive Accident Plan focused exclusively on minor incidents, assuring its readers that "the possibility of a severe accident occurring is so small that, from an engineering standpoint, it is practically unthinkable" (Lochbaum et al. 2014, 16).

States are confident in this belief because their experts have assessed new reactor designs and determined them to be ultrareliable. (Westinghouse's new AP1000 reactor, for example, has been calculated to enjoy a "core melt frequency" of no more than once in every 2.4 million reactor-years, and a "large release frequency" of once in every 27 million reactor-years [Sutharshan et al. 2011: 297–298]). And states are comfortable trusting such determinations because they see no a priori distinction between claims about ultrahigh reliabilities in reactors on one hand, and myriad other engineering claims of more demonstrable efficacy on the other, not least those made about the reliability of jetliners, which are ostensibly equivalent and undeniably well grounded.

It is common, in fact, for jetliners to be invoked in discussions of reactor safety in ways that imply that the technologies pose comparable challenges regarding their governance (e.g., Bier et al. 2003). To a greater degree than with most catastrophic technologies, therefore, it is worth considering their relationship in more detail.

14.2 JETLINERS AND REACTORS

PARALLEL STRUCTURES
Even more than jetliners, reactors exemplify the approach to governing catastrophic technologies outlined in chapter 1, wherein safety is framed as a function of reliability and reliability is presented as a definitively (and predictively) knowable property of complex systems.[3] It is worth noting, in fact, that this approach actually originated in the US nuclear sphere for the explicit purpose of bounding disasters out of public discourse (Downer and Ramana 2021; Wellock 2017; Rip 1986; Miller 2003a). (To tell this story here would require an excessive detour, but in brief, it was driven by the need to justify the reactor program to Congress, and was fiercely contested by critics at the time [Okrent 1978; RARG 1978].)[4]

What began in the civil nuclear sphere, however, soon colonized civil aviation. The structures and practices through which the US—and, largely as a consequence, most other nations—manages reactors and jetliners have long coevolved, with each influencing the other and both taking similar forms as a result. So it is, for example, that in both spheres, a dedicated regulatory agency promulgates and enforces formal certification standards to ensure the reliability of new designs prior to their operation. The NRC's equivalent to the FAA's type-certification process is its design-certification process, overseen by subdivisions of its Office of New Reactors. Like type certification, it stipulates quantitative failure-performance requirements and assessment practices rather than mandating specific engineering choices, and it relies extensively on delegation to ensure compliance (NRC 2009, 18).

It makes sense that jetliners and reactors would engender similar oversight and governance structures. The two technologies—born within a few years of each other (1952 and 1954, respectively)[5]—appear to pose similar challenges. Both are charismatic, highly complex, tightly coupled sociotechnical systems, designed for near-constant operation over several decades. When in operation, both must actively work to negate powerful countervailing forces (gravity and fission, respectively), and each demands broadly equivalent levels of reliability. (As previously discussed, the consequences of meltdowns are much greater than those of plane crashes, but with many more jetliners in operation than reactors, the reliability required of each is about the same.)[6]

Such parallels have made the nuclear sphere a principal beneficiary of the credibility that civil aviation has lent to the positivist portrayal of ultrahigh reliability management. It is ironic, therefore, that reactors exemplify the ways in which most catastrophic technologies are radically different from jetliners with respect to their reliability.

UNDERLYING DIFFERENCES

Ostensibly very similar when viewed through the lens of rule-governed objectivity, reactors and jetliners look very different when understood in relation to the principles and practices outlined in preceding chapters as being essential to ultrahigh reliability. Perhaps the most fundamental differences in this regard relate to their cumulative service experience. As we have seen, reactors operate in much smaller numbers than jetliners. Tens of thousands of jetliners traverse the skies every day, whereas the World Nuclear Association (2019) was claiming just 450 reactors in operation as of 2019, spread across plants in thirty countries.

Compounding this difference in absolute numbers, moreover, are dramatic design variations between reactors, which greatly reduces the relevance of any one reactor's service data to most others. "Similarity" might be impossible to quantify precisely, but even the most cursory taxonomy of reactor architectures makes it clear that the nuclear industry does not practice design stability in anything like the same fashion as civil aviation. At the very broadest level, for instance, the world's reactors can be categorized into a range of types, including pressurized water reactors, boiling water reactors, pressurized heavy water reactors, gas-cooled reactors, and molten salt reactors, which vary even in their fundamental operating principles. Each type can further be divided into multiple generations. (The industry divides pressurized water reactors into four generations, for instance.) And even reactors of the same type and generation are highly variable, partly because there are dramatic variations among the offerings of different manufacturers, but also because each reactor is tailored to suit specific local conditions (such as seismic requirements), a process that creates significant variations between otherwise identical designs (IEE 2005).

These differences in operating volume and design stability mean that the civil nuclear industry's statistically relevant service experience is radically far than that enjoyed by its civil aviation peers. In 2019, again, the World Nuclear Association (2019) was claiming "more than 17,000 reactor-years of experience," a number that pales in comparison to the 45 million-plus years of experience that jetliners accrue every single year, especially given the manifest dissimilarities among the reactors it is counting. This difference in service experience, in turn, dramatically shapes the nuclear industry's relationship to ultrahigh reliability, because "17,000 reactor years of experience" doesn't mean much when modern reactors are expected to melt down less than once in every million years. Even if we imagined that all reactors were identical in their designs, and their years of operation were completely failure free—neither of which is remotely true—that data would be insufficient for statistically testing the industry's reliability claims.

In practical terms, this relative dearth of experience means that experts in the nuclear sphere have not been able to hone reactor designs (and assessments of those designs) in anything like the same way as their counterparts in civil aviation. Modern jetliners have proved that in principle, engineers can leverage the hard-earned lessons of experience to gradually achieve extreme reliabilities in complex systems, given enough time and enough design stability. But engineers working in the nuclear sphere enjoy neither of these resources in anything close to the required amount. The limited number of reactors in operation offer far fewer lessons from which they might learn the limits of their abstractions, and the diversity of reactor designs limits the relevance (and thus the usefulness) of any lessons that they do learn.[7]

The nuclear industry's paucity of service experience also has implications for its incentive structures, which again differ dramatically from those that frame the choices made in civil aviation. With so few reactors in service, those reactors could all be far less reliable than promised and still operate for an extremely long time (probably their entire service lives) before that unreliability manifested in a catastrophic failure. Even though rare, such failures would still be intolerable, given their potential consequences, but the odds of them affecting any specific manufacturer, regulator, or operator would be low. And in the unlikely event that an organization *did* see a catastrophe in one of its reactors, the odds are good that many years would have passed since the failed reactor had been commissioned, designed, approved, purchased, and built. This would mean that the individuals involved in these processes would have moved on to new roles (or retired), and the discourse around reactor safety would be tied to more modern and substantially changed designs.

In combination with other factors—such as statutes that indemnify operators against catastrophic losses[8]—these conditions dramatically limit the costs that organizations can incur for falling short of their extreme reliability assertions. In doing so, they structure the landscape of incentives that act on those organizations. It is easy to imagine this incentive structure weighing heavily on organizational attitudes, cultures, and interpretations across the industry, especially given the longstanding political and economic precarity of atomic energy, which keeps all its associated investment and employment in near-constant jeopardy.

For an insight into this dynamic, we might look to an introspective article by a nuclear engineering consultant in the aftermath of the Chernobyl accident in 1986 (Davies 1986, 59). He writes:

> At some point in their careers most nuclear engineers and scientists have had to resolve in their own minds the awful dilemma posed between the potentially catastrophic effects of a serious accident at a nuclear power plant and the remoteness

of the possibility of its occurrence. Most have squared the circle by concluding, consciously or otherwise, that the remoteness of the chance of an accident was such that serious accidents likely *would* never happen and even *could* never happen. Even if one did happen, it would not be at your plant.

FUNDAMENTALLY UNSAFE

This is all to say that any symmetries between the reliability problems posed by jetliners and reactors are superficial and misleading. When extreme reliability is understood as depending on things like consecrated service experience and favorable structural incentives, as well as just rigorous analysis, then it becomes clear that these technologies are not alike.

Reactors are extraordinary engineering achievements, built and assessed by accomplished and dedicated experts. Nobody can deny this. Yet their failure behaviors will always be subject to epistemological and organizational constraints that no amount of effort or analysis could hope to overcome. Making reactors as reliable as jetliners would require the industry to commit to a common design paradigm, it would mean building tens of thousands of nuclear plants, and it would mean enduring many more accidents like Chernobyl and Fukushima. The first two of these conditions are unrealistic on many levels—even if they were incentivized, which they are not—and the last is intolerable; the costs of nuclear accidents will always eclipse the value of any insights they could offer.

Absent these conditions, the experts who design and evaluate reactors are wholly dependent on their abstractions. And the extreme behaviors they are representing—failure performances over billions of hours of operation—demand an unrealistic degree of perfection from those abstractions. Over enough time, therefore, it is reasonable to expect unforeseen failures arising from fundamental ambiguities in the knowledge on which reactors are built—what I have called "rational accidents."

For evidence that the reliability claimed of reactors is indeed unrealistic, we need look no further than the industry's service record. Statistically, proving ultrahigh reliability requires a huge amount of failure-free operation, but *disproving* it only takes a tiny number of catastrophic failures. This means that the history of nuclear energy could be dispositive, if only we allowed it to speak.

The exact number of historical reactor accidents varies depending on the definitions of key variables—whether Fukushima should count as one "plant" meltdown or as three separate "reactor" meltdowns, for example[9]—but relatively conservative estimates put the rate of serious accidents at somewhere between 1 in every 1,300–3,600 reactor-years (Ramana 2011; Taebi, Roeser, and van de Poel 2012, 202–206; Suvrat 2016). Raju (2016, 56) offers an uncommonly sophisticated analysis of this question and comes to an unambiguous conclusion: "It is clear that the results of [nuclear safety assessments] are untenable in the light of empirical data."

Yet we do not allow this record to speak. With very few exceptions, public accounts of reactor accidents interpret them as evidence not of fundamental limits to engineering ambition but of avoidable human failings: correctable aberrations with limited implications for reactor safety more broadly. Chernobyl, for example, is widely remembered as a Soviet accident more than a nuclear accident—the result of failings specific to regional design, oversight, and operating practices, rather than anything more generalizable (Socolow 2011).

Fukushima is understood in much the same way. Official and unofficial accounts of the disaster evince a wide range of narratives. Some highlight specificities of the plant's design or geography, for example, while others point to failures in Japan's specific regulatory system or corporate governance (Downer 2014; Hamblin 2012). Few, if any, frame the accident as evidence that ultrareliable reactors, as well as accurate assessments of those reactors, are fundamentally unachievable goals premised on unrealistic expectations about the knowledge on which they are built. Even though these narratives often highlight real failings, therefore, they routinely miss the larger point: that unavoidable catastrophes lurk in the irreducible uncertainties of their design and operation.

This is a shame, as the story of Fukushima illustrates the point about irreducible uncertainties quite well.

14.3   FUKUSHIMA'S FAILING

FUNDAMENTAL ERRORS
In the most basic and fundamental sense, Fukushima failed because it was built on erroneous assumptions. As with all nuclear plants, its design was framed by a broad set of premises, codified in a set of documents referred to as its "design basis." At the most fundamental level, the design basis of a nuclear plant consists of two interrelated elements. The first is an understanding of

the parameters within which the plant must operate: for example, environmental conditions and any hazards it might face. The second is a set of technical requirements that reflect those parameters: for example, the minimum tensile strength of specific structural elements or the minimum separation distance between certain subsystems (Wyss 2016, 31).

So it was that, among many other things, Fukushima's design basis included specifications relating to potential tsunamis: specifying the hazards they posed and how those hazards were to be neutralized. Specifically, it set the maximum height of any tsunami that the plant could face at 3.1 meters (about 10 feet); and to ensure a comfortable margin of safety, it required the plant to have flood defenses capable of withstanding a tsunami almost twice that height: 5.7 meters (19 feet). These measures, its designers and regulators decided, would reduce the risk of tsunami inundation to essentially zero.

This conclusion, that tsunami inundation was functionally impossible, then shaped numerous downstream design and planning choices. Fatefully, for example, it informed a decision to place all the plant's redundant backup generators underground for maximum protection from terrestrial hazards. It also influenced the siting of the plant, the elevation of which engineers lowered by 25 meters (80 feet) so that the reactors would stand on bedrock and be more resilient to earthquakes (Lochbaum et al. 2014).

In the event, however, the tsunami that swamped Fukushima in 2011 far outstripped the levels specified in its design basis. At somewhere between 12 and 15 meters (40–49 feet)—over four times the calculated maximum—it crashed over the plant's seawalls, swamping the entire site and drowning all the underground emergency generators simultaneously in an unanticipated "common mode" failure. Thus began a cascade of failures that quickly cumulated in catastrophe. Faced with a massive regional power outage, the plant was unable to fall back on its nonfunctioning generators to cool its furiously heat-generating reactors. Untempered fission then did as untempered fission does, and the system inexorably spiraled out of control.

Viewing Fukushima's failures in this way—as born of a foundational error in the plant's design basis—muddies narratives about flaws in its design. The disaster's immediate wake saw various embattled industry proponents gamely contending that the plant had not actually failed at all, since it had never been designed to survive the conditions that brought it down (e.g., Harvey 2011; Hiserodt 2011). The American Nuclear Society (2011) even went so far as to argue that Fukushima "could actually be considered a 'success,' given the scale of this natural disaster that had not been considered in the original design." The wider logic of such claims is contestable at best—the plant clearly failed, by any reasonable definition of the word—but the frustration is understandable. The plant's design basis was established at the earliest stages of its conception. Its underlying assumptions about tsunamis must have been all but invisible to the experts who then used it to plan, build, and assess the plant itself. No additional effort on their part would likely have revealed the error. They extensively modeled and tested the plant's design, and those models and tests—which were framed on the same foundational belief about tsunamis—repeatedly validated their conclusions. The plant's superintendent spoke for almost everyone involved in its safety when he told reporters that "we encountered a situation we had never imagined" (Lochbaum et al. 2014, 13).

In light of all this, it is tempting to blame the experts who created Fukushima's flawed design basis for the plant's demise, but this too is problematic. Examined closely, there are few reasons to doubt that the design basis reflected the best understanding of tsunamis that was available when construction on the plant began in 1967. The maximum height calculation was wrong, in essence, because seismologists had substantially underestimated the potential size of earthquakes in the region, but this error would not have been apparent at the time. The Tōhoku megaquake that felled Fukushima was an estimated magnitude—9 or 9.1—that was not even recognized as geologically possible in 1967. "Magnitude 9" only became a generally recognized class of earthquakes around 1980, and for decades afterward, many seismologists continued to believe they were limited to certain types of subduction zones and could not occur near Fukushima. History seemed to confirm this assessment. Prior to the twenty-first century, there was no record of such an earthquake, or a tsunami, ever occurring in the region (Nöggerath, Geller, and Gusiakov 2011, 38–42; Lochbaum et al. 2014, 52).

It is also true, of course, that Fukushima's demise came long after the plant was built, and many critics have pointed out that science's understanding of earthquakes had evolved considerably by 2011. As noted previously, seismologists began to recognize the existence of megaquakes in the late 1970s and early 1980s. And by the early 2000s, new techniques for identifying and interpreting sedimentary rocks had led some "paleo-tsunami" researchers to conclude that at least three such quakes had occurred in the region over the previous 3,000 years (Nöggerath et al. 2011, 41; Minoura et al. 2001). The nuclear community was aware, or at least not universally *unaware*, that these findings had implications for Fukushima Daichi. In the years leading up to the disaster, several bodies had formally questioned the plant's design basis on these grounds. In 2002, for instance, the Headquarters for Earthquake Research Promotion issued revised predictions for the region, which, when modeled by the plant operator (TEPCO), implied a potential tsunami of 51 feet (15.7 meters) at the site (Lochbaum et al. 2014, 52).

Understandably, therefore, several accounts of the disaster contend that the plant should have been updated to reflect this new evidence (e.g., Nöggerath et al. 2011; Acton and Hibbs 2012). Critics point to missed opportunities created by the new seismology, observing, for instance, that Japan's Nuclear Safety Commission revised its seismic guidelines in 2006 without mandating significantly upgraded tsunami protections, and it failed to act in 2009 after a senior geologist issued strong warnings about tsunami risks (Nöggerath et al. 2011, 42). These missed opportunities look damning in hindsight, and many claim they constitute a culpable failing that should have been avoided. By this view, Fukushima might have been blameless in 2000, but not by 2011.

In principle at least, the design bases of nuclear plants are living documents that can be revised over time, so the claim that they should have been updated is not inherently unreasonable. Still, however, the "failure to heed new seismology" critique is unsatisfying on at least two significant levels. The first is that it assigns blame but offers no comfort, for even if we allow that Japan's nuclear establishment should have revised or closed Fukushima in light of new seismology, then there was still a period of several decades during which there was no reason to doubt the flood defenses. The accident could easily have occurred in the years prior to the discovery of megaquakes, in other words. And, had that occurred, the plant's design basis would have been wholly congruent with the best science available. Those who claim that Fukushima is blameworthy on these grounds, therefore, are also tacitly conceding that failures—rational accidents—are sometimes inherently unavoidable. When contemplating disasters that might threaten major cities, this is not a reassuring thought.

The second issue with this critique is that it isn't very compelling even on its own merits. This is because hindsight gives a misleading perspective on the science and the way it was received. In debating whether Fukushima's design basis should have been altered, it is important to understand that the new seismological findings were vigorously contested within the expert community itself. TEPCO and its regulator countered them with rival findings, also backed by respected seismologists, based on models and simulations that either placed less emphasis on historical earthquake data (e.g., Yanagisawa et al. 2007) or framed that data in ways that downplayed its tsunami implications (Nöggerath et al. 2011, 42). As in all such debates, moreover, policymakers were ill equipped to judge the relative merits of different positions (Hirano 2013, 18–19). "[C]onfusion reigned in the field of seismic risk," as Lochbaum et al. (2014, 44) put it, with "even state-of-the-art science [being] unable to shed much light on the question of 'how safe is safe enough' when it came to building nuclear power plants in Japan."

So it is that, while the nuclear community's resistance to new estimates of seismological risk was undoubtedly self-serving, and perhaps even contrary to the broader scientific consensus, it was at least defensible. Every new scientific finding is a convoluted and incremental achievement, especially if they materially affect powerful institutional actors (Jasanoff 1990; Salter 1988). Complex research is never accepted promptly by parties that stand to lose from it, and in this instance, the nuclear community (and the state itself) stood to lose a lot. It is doubtful whether it would even have been possible to update Fukushima Daiichi to defend against earthquakes and tsunamis that were substantially more severe than those outlined in its original design basis.

In a world where some major polities still barely recognize the existence of anthropogenic climate change, therefore, it is wholly unrealistic to imagine that a private corporation would promptly condemn, or even substantially rebuild, a nuclear plant when confronted with esoteric and contested new paleo-tsunami findings, or that a state would require it to do so. (It is worth noting that the US, for its part, had done little to revise its reactors as seismology evolved [Dedman 2011; Lochbaum et al. 2014, 115].)[10] Arguments that invoke this expectation to assign blame for the Fukushima disaster are probably better understood as critiques of modernity itself.

Insofar as this expectation is unrealistic, moreover, then Fukushima might reasonably be understood as a rational accident. Seen in this way, its wreckage and fallout are a monument to technological hubris; they represent a failure to see the limits of what engineers can know, and what those limits imply for our technological ambitions.

NOTES

INTRODUCTION

1. The term reliability is being used narrowly here. As, more conventionally, it might also encompass non-catastrophic failures.

2. It is worth noting that Vaughn (2021; 1996) is unusual in having authored signal texts from both sides of Sagan's divide. There is nothing inherently problematic about this, of course, but it does suggest that we should be cautious of categorizing scholars themselves, as commonly occurs.

3. Insofar as a principled disagreement does exist in the scholarship on this question, it probably lies in whether technologies like reactors and atomic weapons can ever be "safe enough": a question that is as much about morality and politics as it is about safety practices.

4. I tried to make this point in an earlier publication (Downer 2011b), but I framed it more as a fundamental difference of principle, which was an error on my part.

CHAPTER 1

1. Take, for instance, estimates of the accident's financial costs. In 2013, these were usually pegged at around $100 billion. By 2017, however, the Japan Institute for Economic Research was putting them in the region of $449–628 billion, simply from decommissioning, decontamination, and compensation (Burnie 2017). This figure does not include indirect costs arising from the accident's massive disruption to the country's energy sector, or wider losses to tourism, agriculture, and industry.

2. In 1959, the RAND Corporation published a top-secret report outlining how the US should manage the publicity fallout from an accidental nuclear warhead detonation. It suggested that authorities aggressively delay and draw out the release of

damaging information because the media's interest would inevitably wane over time (Schlosser 2013, 195). Whether by design or by chance, information about Fukushima has followed a similar trajectory, with RAND's predicted outcome.

3. The "banana equivalent dose" is an informal measurement of radiation exposure premised on the fact that bananas contain radioactive potassium. It is favored by some industry communicators, especially those eager to downplay radiological hazards, but it obfuscates more than it illuminates because it ignores the fact that humans excrete excess potassium (WashingtonsBlog 2013; EPA 2009, 16).

4. The language around this promise can be misleading. Strictly speaking, expert bodies (at least in the US) rarely assert that failures in acute catastrophic technologies are impossible. Such claims are too difficult to defend. Instead, those bodies tend to assert that failures in such systems are too improbable to take seriously, despite their potentially dire consequences. For almost all practical purposes, however, the claims are equivalent. The levels of reliability demonstrably achieved in jetliners, when applied to systems built in much smaller numbers, imply a vanishingly small risk of catastrophic failure. And states have long been comfortable ignoring risks they deem possible but implausible. Few would deny that extra-terrestrial invasions or apocalyptic asteroid strikes are hypothetically possible, but neither contingency is afforded much serious (or even superficial) deliberation.

5. The relative nature of aviation safety is exemplified by a 1920 book titled Aerial Transport, which boasted that "[t]he argument that flying is inherently dangerous cannot in fact be substantiated," citing data showing that 40,000 miles were being flown between fatal accidents (quoted in Chaplin 2011: 78). The same rate today would imply well over a million fatal accidents every year.

6. As of this writing, there are about sixty-five times as many jetliners as there are reactors, and their operational downtimes are more equivalent than is probably intuitive. So, as an extremely rough rule of thumb, if each failed as frequently as the other, then we might expect about sixty-five plane crashes for every reactor meltdown.

7. Atomic weapons are unique in the sense that there are potentially extreme dangers that arise from not building them, as well as from them failing, which complicates any societal risk calculus involving their failure behavior. This tension is evident in Sagan (1993).

8. The NRC and the US Department of Defense, for instance, both explicitly recommend the use of use of qualitative "human performance reliability analysis" in their assessment guidance for critical systems (e.g., DoD 2000, 1999; NRC 1983).

CHAPTER 2

1. Efforts to quantify failure performance predate this shift. In the early 1800s, for instance, railway development inspired extensive studies on the life spans of roller-bearings (Villemeur 1991). In the 1920s, meanwhile, Bell Laboratories, struggling with

the mercurial performance of vacuum tubes, formed a "quality" department that pioneered various quantitative reliability metrics (Zimmel 2004; Shewhart 1931). It was not until World War II, however, that such techniques started to become standard engineering tools. And it was primarily during the Cold War that mathematicians and engineers began to develop the conceptual tools, metrics, and categories that normalized reliability as a quantitatively expressible and manipulable entity. It was also at this time that reliability engineering emerged as a distinct specialty within the discipline and profession (Jones-Imhotep 2002; Regulinski 1984; Kececioglu and Tian 1984).

2. "Service data," in this context, might loosely be understood as referring to the combination of the number of systems in operation multiplied by the duration of their operation. For example, 100 identical systems, each operating for 100 hours, will accrue 10,000 hours of service data, and so will 1,000 identical systems, each operating for 10 hours. As we will see, however, this equivalence is not perfect.

3. Such *ceteris paribus* assumptions are not always straightforward, however. When the US infantry started fighting in Vietnam, for instance, much of the reliability data for its rifles was rendered moot by unexpected operational practices and the humid operating environment (Fallows 1985).

4. Lynch and Cole (2002), for instance, make this argument in relation to the reliability of DNA testing. They point out that the oft-cited one-in-a-billion chance of a false positive is nonsensical, given that the odds of the test being contaminated or tampered with is significantly higher than this.

CHAPTER 3

1. The Airbus A380 carries up to 320,000 liters (roughly 84,500 gallons or 260 tons) of fuel.

2. As of 2017, reportedly the oldest passenger aircraft still in regular service was forty-seven years old. It was a Boeing 737-200, serial number 20335, flying domestic routes for Airfast Indonesia (Smith 2017).

3. These jetliners were operated by 1,397 individual airlines, in 3,864 airports, serving 49,871 routes and carrying 2.97 billion passengers during the course of each year.

4. This claim admittedly deserves some caveats. There were nonfatal jetliner incidents, such as Air France Flight 66, an A-380 that suffered an uncontained engine failure and made an emergency landing in Canada. There was a fatal jetliner cargo flight: Turkish Airlines Flight 6491, a B747 that crashed in Kyrgyzstan while trying to land in thick fog, killing four crew members and thirty-five people on the ground. There were also a few accidents with turboprop airliners (as opposed to jetliners), such as West Wind Aviation Flight 280, which crashed while taking off in Canada, killing one passenger.

5. The last US fatal accident prior to 2018 had been Colgan Air Flight 3407, which crashed on approach to Buffalo on February 12, 2009. The streak was sadly undone

on April 17, 2018, when Southwest Airlines Flight 1380—a Boeing 737 en route to Dallas from New York—suffered an engine failure that peppered the fuselage with shrapnel, killing one passenger.

6. The data tell the same story when broken down by specific airframes. Take, for example, the Boeing 747, a once-popular airframe, but by no means the most common. Introduced in 1970, Boeing had built almost 1,500 747s by 2010, which had made over 17 million flights, covering a cumulative distance of over 42 billion nautical miles. During that period, they were involved in only thirty-four separate incidents resulting in multiple fatalities (Airsafe 2010).

7. As already discussed, the fact that jetliner crashes are less consequential than (for instance) reactor meltdowns is offset, in terms of the reliability they require, by the fact that there are many more airplanes than reactors.

CHAPTER 4

1. Engines and airframes are meaningfully distinct in this context. They are manufactured and maintained by different companies and regulated under separate codes (albeit codes governed by very similar principles). Airlines also purchase them separately from the airframes themselves, there usually being multiple engine options for any given airframe.

2. These offices—located in Seattle, Denver, and Los Angeles—are overseen by the FAA's Aircraft Certification Service through its Transport Airplane Directorate based in Renton, Washington. As of 2006, the Transport Airplane Directorate and its three Aircraft Certification Offices employed about 250 technical people to assist in certification activities (NTSB 2006, 68).

3. Prior to digitization, at least, this was literally true. As early as 1980, Lockheed was estimating that in the course of certificating a new Jetliner, it would submit approximately 300,000 engineering drawings, 2,000 engineering reports, 200 vendor reports, and about 1,500 letters (NAS 1980, 29).

4. This is a simplification, albeit a common one. Title 14 of the Code of Federal Regulations actually includes other rules applicable to certifying a transport-category airplane. These include Part 21, "Certification Procedures for Products and Parts"; Part 33, "Airworthiness Standards: Aircraft Engines"; Part 34, "Fuel Venting and Exhaust Emission Requirements for Turbine Engine Powered Airplanes"; and Part 36, "Noise Standards: Aircraft Type and Airworthiness Certification."

5. Advisory Circulars explain acceptable ways to comply with certification requirements. Notable among them for the argument that follows is "AC 25.1309-1: System Design and Analysis," which outlines the rationale for FAR-25's reliability requirements and explains, in broad terms, how manufacturers should demonstrate a system's compliance (FAA 1988, 2002a, 2002b).

6. So it is, for example, that FAA Advisory Circular 25.1309-1 has an appendix called "Background Information for Conducting Failure Analyses," which directs readers to "NUREG-0492," an NRC publication, and "MIL-HDBK-217," a Department of Defense standard.

7. Prior to this, US civil aviation operated without specific federal oversight. In principle, a purblind inebriate could fly loops over Times Square without breaking any specific laws. Perhaps surprisingly, calls for federal regulation were largely driven by the fledgling airlines themselves. They saw regulation as essential to commercial success, in part because more established and trusted modes of public transport, like ferries and railways, had long been subject to oversight (Hansen, McAndrews, and Berkeley 2008). Their argument was bolstered by the experiences of the US Post Office. The Air Mail Service—by far the most significant US operator of aircraft at the time—had implemented its own oversight measures, and these were demonstrating their value. It was averaging one fatality every 463,000 miles, compared to one every 13,500 miles for other commercial flights (Nader and Smith 1994, 4).

8. At first, this was the Aeronautics Branch of the Department of Commerce, created by the 1926 Air Commerce Act. The Aeronautics Branch was reestablished as an independent unit, the Federal Aviation Agency, in 1958, and in 1966, it was subsumed under the Department of Transport and renamed the Federal Aviation Administration (Briddon et al. 1974; Cobb and Primo 2003, 16). The FAA remains under the department's aegis today, although it largely works autonomously.

9. Most of the systems in an aircraft (essentially anything that draws electricity) are assessed almost entirely probabilistically, which is to say that experts calculate their reliability directly to ensure that it satisfies requirements. Structures (such as wings, windows, and fuselage sections), by contrast, are effectively still governed by proscriptive design requirements (FAA 1999a). The process calculates the reliability of these elements indirectly, incorporating them into reliability assessments of the integrated airframe by assuming that, by virtue of meeting specified design requirements, they have a preestablished failure probability (usually zero). The distinction between "structures" and "systems" is important, therefore, but often opaque. "Fly-by-wire" systems, for instance, integrate complex avionics with structural control elements that were traditionally governed through different processes. The NTSB has argued that ambiguity around this distinction has allowed catastrophic weaknesses to fall through cracks in the assessment process. (Take, for example, Alaska Airlines Flight 261, which crashed off the coast of California in January 2000 because of a stabilizer failure. Investigators apportioned some of the blame for this to the certification analysis, which had treated the failed part as "structure" rather than a "system" [NTSB 2000b; 2006, 18, 52–60]).

10. These tools are represented by a fruit salad of acronyms. They include Failure Mode and Effects Analysis (FMEA), Failure Modes and Effects Summary (FMES), Functional Hazard Assessment (FHA), Preliminary System Safety Assessment (PSSA), System Safety Assessment (SSA), Fault Tree Analysis (FTA), Common Cause Analysis (CCA), Zonal

Safety Analysis (ZSA), Probabilistic Risk Analysis (PRA), and Common Mode Analysis (CMA). Many of these, and others, are outlined in Aerospace Recommended Practice ARP4761 (SAE International 1996).

11. To complicate matters further, the operational unit of certification might also be construed as the airplane's "catastrophic failure modes." There is an ambiguity in the guidance between "catastrophic failure modes" and "flight critical systems." Organizationally, the certification process is structured around "critical systems." Confoundingly, however, the ultimate rationalization for the reliability required of those systems is often expressed in terms of "catastrophic failure modes." In these instances, the regulations do not state the maximum frequency of critical system failures but the minimum frequency of catastrophic airplane-level failure conditions (e.g., a loss of power) (FAA 2002, 10–11). The two are routinely equated, however, such that having 100 catastrophic failure modes is understood to imply 100 critical systems. (Thus, this assumes a direct one-to-one relationship between the two that is itself problematic.) Attempting to parse the close logic of certification can be trying, but this is the point.

12. Critical systems are identified in relation to the failure modes that they would create if they failed. So it is, for instance, that redundant elements that would have to fail simultaneously to create a catastrophic failure are grouped together as a single system.

13. This understanding about the relationship between the reliability of individual components to that of the wider system has its roots in the V2 rocket program. Germany's rocket engineers, who struggled enormously with failure, had started out using reliability principles based on a simple and then widely held understanding that "a chain was as strong as its weakest link." They were forced to abandon this assumption when tests proved it to be wildly inaccurate, however, and after consulting the mathematician Eric Pieruschka, concluded that the reliability of a system was actually the product of the reliabilities of its constituent elements. (Thus, the reliability of a system composed of identical elements would be $1/x^n$, where x is the reliability and n is the number of elements [Villemeur 1991, 5].) This, in turn, led to the conclusion already given—namely, that the reliability of individual components and subsystems must be much higher than the desired reliability of the system.

14. This difference is not immediately obvious in the text, which actually states that "a fleet of 100 aircraft of a type, each flying 3000 hours per annum, one or other of the Catastrophic Effects might be expected to turn up once in 30 odd years, which is close to the concept of 'virtually never'" (Lloyd and Tye 1982, 37). (This amounts to 3,000 hours per annum $\times$ 30 years = a service life of 90,000 hours.)

15. Anonymous correspondence (March 25, 2005).

16. Anonymous correspondence (March 25, 2005).

17. Anonymous correspondence (March 24, 2005).

18. The guidance appears to have gradually accrued such admonitions and caveats since its initial turn to probabilism. In part, they reflect years of misgivings about quantitative calculations voiced by the NTSB, often in accident reports that have drawn attention to unrealistic and insufficient numbers being used in certification assessments. (After TWA Flight 800 exploded outside JFK Airport in 1996, for example, the NTSB found that "undue reliance" had been placed on formal risk tools when calculating the probability of a fuel vapor explosion, and that such tools "should not be relied on as the sole means of demonstrating [compliance]" [NTSB 2000a, 297]). More directly, however, they reflect protracted debates that arose when the certification process first began to incorporate software (e.g., FAA 1982, 9; 2002, 7).

CHAPTER 5

1. The US Air Force, which routinely flies at lower altitudes, incurs commensurately greater costs relative to its flight hours. In 2003, it estimated that bird strikes were responsible for killing two of its aircrew and downing two of its aircraft every three to five years, costing the service between $50 and $80 million every year (Feris 2003).

2. Birds are poor at avoiding fast-moving objects because they react to proximity regardless of velocity. They usually take evasive action when incoming threats are about 100 feet away: a strategy that works well for them until those threats are traveling at more than about 55 miles per hour. Scientists established this, in part, by driving pickup trucks at unsuspecting turkey vultures (DeVault et al. 2014; 2015).

3. By contrast, the average hatchback, as one major engine manufacturer is pleased to observe, is barely worth its weight in hamburger.

4. Only the largest engines are subjected to the largest birds because it is thought that large birds entering smaller engines tend to strike the cowling and break up before they hit the blades (FAA 1998).

5. Engines certified since November 2007 have been assessed under revised regulations that include an additional class of bird: the "large flocking bird." Tests for this class involve a single bird (either 4, 4.5, or 5.5 pounds [1.8, 2, or 2.5 kilograms] depending on the engine size) that is directed into the fan blades rather than the engine core and require that the engine maintain some thrust (which differs at specific time points) after impact (NTSB 2009, 19, 84).

6. The density of bird traffic decreases exponentially with altitude. The FAA has determined that 71 percent of bird strikes on commercial aircraft occur below 500 feet (from ground level). Above this height, they decline by 34 percent for each 1,000-foot gain in altitude (FAA 2014, xi). Bird strikes are 1.5 times more common during arrival, but 3.4 times more damaging during departure, when the engines are working harder (Dolbeer 2007, 1).

7. Snarge is swabbed and samples are sent to the Smithsonian Institution's Feather Identification Laboratory for DNA analysis, but interpreting the samples is an imperfect

art. As of 2009, for example, the lab was unable to discriminate between multiple birds of the same species and sex (Budgey, 1999; NTSB 2009, 80fn).

8. Only dead birds are used in the tests. Even if engineers were able—and, it should be said, willing—to coax a live bird into one end of a powerful compressed-air cannon, that bird would certainly be dead upon exiting the barrel at 200 knots. A surprising number of people enquire about this.

9. It is not entirely unusual for scaling variables to have nonlinear effects in engineering (see, e.g., White 2016, 338–339). It is worth noting, however, that the JAA's conclusion has far-reaching implications for the underlying logic of modern ingestion standards, which presume a linear relationship between most variables, including mass and damage. The FAA, for instance, presumes a 1.28 percent increase in the likelihood of damage for every 100-gram increase in body mass (FAA 2014, xi).

10. It is usually considered best if the leading fan blades can slice a bird into something resembling salami before it passes through the rest of the engine. Back in the less euphemistic 1970s, the FAA listed "blades which effectively mince birds upon contact" among its "desirable engine features" (FAA 1970, §3).

11. A total of 52 percent of bird strikes occur between July and October, when birds are gathering to migrate (FAA 2014). The US Air Force, which enjoys an authoritative euphemism, often refers to such gatherings as "waves of biomass."

12. Given the vast legions of chickens that become McNuggets, the handful of birds sacrificed for the betterment of aviation safety seem to elicit an incongruous degree of public anxiety.

13. Engines have failed their bird-strike tests in the past, with meaningful consequences. Rolls-Royce can testify to this. By 1969, the British engine manufacturer had developed, at great expense, a novel fan blade made from an innovative material called "Hyfil." After meeting every other requirement, however, the blades failed their bird-strike tests. The company was forced to abandon Hyfil and return to titanium as a result (Spinardi 2002, 385). This, in turn, made the engines heavier and thus unable to meet Rolls's fuel-consumption guarantees, strongly contributing to its bankruptcy the following year (Newhouse 1982, 174). (It is worth noting, however, that these failures occurred in precertification tests.)

CHAPTER 6

1. "Ladies and gentlemen, this is your captain speaking. We have a small problem. All four engines have stopped. We are doing our damnedest to get them going again. I trust you are not in too much distress," Captain Eric Moody, British Airways Flight 009 (1982).

2. The record was short lived. It would be broken the following year by Air Canada Flight 143, which ran out of fuel mid-flight after ground crew confused imperial and metric units.

3. The eminently quotable captain would later describe the landing as being "a bit like negotiating one's way up a badger's arse" (Faith 1996, 156). From the context, it's clear that he meant by this that it was difficult (the badger presumably objecting to said negotiation).

4. The FAA calls its safety philosophy the "fail-safe design concept." It encompasses a range of principles and rubrics: an emphasis on "margins of safety" is one; "checkability" is another (FAA 2002, 6; NTSB 2006, 86). The plurality of these rubrics depends either directly or indirectly on redundancy.

5. Petroski (1994) argues persuasively for the existence of design paradigms in engineering.

6. Looking beyond aviation for a second, Sagan (1993) reports that false indications from safety devices and backup systems in the US missile-warning apparatus have nearly triggered atomic wars.

7. The type-certification process incorporates the reliability of human beings into its quantitative requirements in a similar fashion to the way that it incorporates that of aircraft structures (as outlined in chapter 4). Designs must meet proscriptive requirements (stipulating minimum cockpit space and sight lines, for instance), which the certification process then assumes will confer mathematically perfect reliability (FAA 1999b). As the FAA (1982, appx 1) puts it, "If the evaluation [of the human system interface] determines that satisfactory intervention can be expected from a properly trained flight crew," and "then the occurrence of the failure condition has been prevented."

8. One example occurred in March 1994, when the pilot of Aeroflot Flight 593 fatefully—and, in the event, fatally—passed the controls of a passenger-laden Airbus A310 to his sixteen-year-old son, Eldar (Norris and Hills 1994, 5).

9. Herein lies another way that human behavior can intrude on redundancy calculations: by undermining the independence of various elements. Investigators of a 1983 multiple engine failure on a Lockheed L-1011, for instance, determined that the engines were united by their maintenance. The same personnel had checked all three engines, and on each one, they had refitted the oil lines without the O-rings necessary to prevent in-flight leakage (Lewis 1990, 207–209).

10. Failures of reserve systems—sometimes known as "latent" or "dormant" failures—can be particularly dangerous because they are more likely to go undetected. Following the crash of USAir Flight 427, for instance, the FAA criticized the Boeing 737's rudder control system on this basis. The system involved two slides: one, which usually did the work, and a second, redundant slide that lay in reserve. The regulator argued that since the system rarely used the second slide, it was prone to fail "silently," leaving the aircraft "a single failure away from disaster" for long periods (Acohido 1996).

11. Anonymous correspondence (July 21, 2006).

12. Thus stricken, the aircraft—Air Transat Flight 236, en route to Lisbon from Toronto with 291 passengers—glided for 115 miles before making a high-speed touchdown

in the Azores that wrecked the undercarriage and blew eight of its ten tires (Airsafe 2008). The incident also has an interesting human dimension, in that the pilot is said to have exacerbated the problem by actively transferring dwindling fuel reserves to the leaking engine.

13. "In most cases," the FAA (2002b, 21) writes, "normal installation practices will result in sufficiently obvious isolation . . . that substantiation can be based on a relatively simple qualitative installation evaluation." The result, as the National Academy of Sciences (NAS 1980, 41) writes, is that "[t]he failure of a neighboring system or structure is not considered within a system's design environment and so is not taken into account when analyzing possible ways a system can fail."

CHAPTER 7

1. This is, by necessity, a slight simplification of a contested and multifaceted engineering explanation, albeit hopefully not a distorted one in respect to the central argument it is being used to illustrate. A fuller account—including a more recently proposed "fluid hammer" hypothesis (Stoller 2001)—would complicate the narrative without changing the underling point.

2. The discipline of structural engineering has a longstanding interest in crack tolerance for this precise reason. "Where human life is concerned," as one classic text on the subject puts it, "it is clearly desirable that a 'safe' crack should be long enough to be visible to a bored and rather stupid inspector working in a bad light on a Friday afternoon" (Gordon 2018 [1991], locs. 1407–1408).

3. The aviation sphere is no exception to this impulse, as is evinced in an oft-excerpted passage from the International Civil Aviation Organization's (ICAO's) first *Accident Prevention Manual*. It states, "The high level of safety achieved in scheduled airline operations lately should not obscure the fact that most of the accidents that occurred could have been prevented. . . . This suggests that in many instances, the safety measures already in place may have been inadequate, circumvented or ignored" (ICAO 1984, 8).

4. The theory rests on the observation that no complex system can function without small irregularities: spilt milk, blown fuses, stuck valves, and so on. Eliminating such irregularities entirely would be impracticable, so engineers construe them as a fundamental design premise: "failure-free" systems being defined by engineers as systems designed to accommodate the many vagaries of normal operation, rather than as systems that *never* experience operational anomalies.

5. It is tempting to see NAT as a restatement of Murphy's Law—the old adage that anything that can go wrong, will—but it is more profound than that. It is a systems-level rethinking of what it means for something to "go wrong," in that it understands failure in terms of a system's organization. Uniquely, it argues that some failures arise from the very fabric of the systems themselves: less a flaw than an emergent property of their underlying structure (Downer 2015).

6. A different reading of NAT, also supported by the text, construes normal accidents as the product of an organizational irony. Grounded in organizational sociology and drawing on contingency theory, it argues that complex and tightly coupled systems call for contradictory management styles: complex systems requiring decentralized management control (because it takes local knowledge and close expertise to monitor and understand the interactions of a diverse network of elements); and tightly coupled systems requiring centralized control (because systems that propagate failures quickly must be controlled quickly, which requires strict coordination and unquestioned obedience) (Perrow 1999, 331–335). This construal of NAT is "weaker," in the sense that it might, in principle, be resolved if sociologists could identify an organizational framework that reconciled the competing demands of complexity and coupling (e.g., LaPorte and Consolini 1991).

7. "Tensile strength" is a metal's resistance to being pulled apart, usually recorded in pounds per square inch. "Elasticity" is a metal's ability to resume its original form after being distorted.

CHAPTER 8

1. A similar concession was voiced by the deputy director of the FAA's Aircraft Certification Division, who said of the standard that governed avionics software (RTCA DO-178A) that it "recognizes that you can't test every situation you encounter" (Beatson 1989b).

2. Flight protections are also thought to have confused pilots from time to time, being implicated as contributory factors in a few accidents (such as Air France Flight 296 in 1988 [Macarthur and Tesch 1999]). Occasions where they have prevented errors go unrecorded, however, although Langewiesche (2009b) makes a compelling argument that they helped save the "Hudson Miracle" flight.

3. This process would be analogous to that described by Galison (1987) in relation to scientific experiments, wherein he argues that scientists' theories about the world become increasingly useful and veridical over time.

4. Anonymous interview. March 29, 2005.

5. Sociologists have observed the same process of rule refinement in other technological spheres. Wynne (1988, 154), for instance, calls the process of rule-making in technological systems "an ever-accumulating practical craft tradition," which he compares to case law.

6. It is also important in this regard that civil avation accidents were always contained in scope: tragic and costly, but never on a scale that could threaten cities, economies, or more than a few hundred people. (Recall from chapter 1 that jetliners might be thought of as a chronic catastrophic technology, in the sense that their extreme reliability requirements owe as much to the numbers in which they are operated, as they do to the extreme hazards of them failing.)

7. Personal communication, May 4, 2005.

8. One example was the so-called airframe revolution, which saw the introduction of retractable landing gear, aluminum stressed-skin structures, wing flaps, and other innovations that became standard.

9. Engines, which are manufactured and purchased separately from the airframe, have followed a distinct but similarly restrained design trajectory—undergoing a shift from low- to high-bypass turbofans in the early 1970s, whereupon they became much wider.

10. Anonymous correspondence, March 20, 2005.

CHAPTER 9

1. Concorde has an uncommon nomenclature, in that it does not receive an article in common English usage (i.e., it is not referred to as "*The* Concorde").

2. It should be said, however, that the Central Intelligence Agency (CIA) was arguably the true pioneer here. The SR-71 was derived from the A-12, which had been developed for the CIA two years earlier.

3. Manufacturers of much smaller, private aircraft could afford to move more quickly. In 1989, for instance, Beech Aircraft Corporation (later Beechcraft) delivered its first Starship: a six-to-eight-passenger turboprop built almost entirely from composites (Huber 2004; Warwick 1986).

4. Flight 961 was forced to land abruptly in March 2005 after losing its composite rudder while high above the Florida Keys. No lives were lost, but investigators found that it had been a close call. The rupture had dangerously stressed the tail itself, the loss of which would have been unsurvivable (TSB 2007). The formal investigation into the incident was not completed until 2007, but long before the report's publication, the industry concluded that the rudder probably had had an undiscovered stress fracture. To prevent its reoccurrence, Airbus issued a directive requiring operators of A300-series airplanes with composite rudders to inspect their outer surfaces for latent damage with an "acoustic tap test" (NTSB 2006, 2).

The incident did not end there, however. In a twist that speaks to the inescapable dangers of epistemological uncertainty, the problem was considered resolved until eight months later, when an unrelated but serendipitous second incident raised new questions about Flight 961's underlying cause. An accidental blow to the rudder of a FedEx A300-600 led maintenance engineers to examine it more carefully than would usually be required. Their examination revealed a shocking amount of damage, but not from the blow. It turned out that, prior to the accidental insult, hydraulic fluid had been leaking from the plane's control system and attacking the rudder's composites. The materials had deteriorated to a point where the structure might easily have failed catastrophically midflight (TSB 2007; Marks 2006). Since the hydraulic fluid had affected the inner rather than the outer surfaces of the rudder, moreover, its damage would not have been revealed by standard test protocols, implying that Airbus's safety inspection program (with its tap tests) was inadequate and its airplanes had all been in danger.

The incident had all the hallmarks of a rational accident born of innovation-driven uncertainty. The implications of hydraulic fluid leaking into composite structures went unrecognized because experts never considered the interaction as a possibility (although the basic science was understood). Tests and models had not revealed the danger because nobody had thought it a relevant variable to test or model; and maintenance practices were not primed to look for the kind of damage that it caused for the same reason (TSB 2007; Marks 2006). As with most rational accidents, moreover, it was informative, spurring a joint NTSB-Airbus investigation into fluid contamination, which informed new maintenance tools and revised procedures (NTSB 2006a).

5. Interestingly, the problematic reliability of military airframes plays an important role in the logic around arms sales. It makes any nation that procures advanced US fighters continuously dependent on the US for spare parts and upkeep, without which their air power would quickly degrade.

6. Reliability problems with military hardware are certainly not exclusive to aviation. The DoD, which leans heavily on technological advantages to achieve its strategic goals, has long struggled with the reliability of its systems. This first became apparent in the early years of the Cold War, when it was discovered that a remarkable portion of the department's expensive, high-tech paraphernalia was down at any given time, and the resources required to diagnose and repair faults was becoming unsupportable. Logistic data from this period testify to the scale of the problem. By 1945, for instance, the navy was annually supplying a million replacement parts to support 160,000 pieces of equipment, and by 1952, it was spending around two dollars every year to maintain each dollar's worth of technology (Coppola 1984, 29). The US defense establishment was acutely concerned by this ever-deepening "reliability crisis," which it saw as jeopardizing its ability to fight wars (Jones-Imhotep 2000, 145).

CHAPTER 10

1. From this perspective, the real mystery of Concorde was not why it disappeared, but why it was built at all. The answer to this question would undoubtedly involve national pride, and the power of what Jasanoff and Kim (2015) would call the Anglo-French sociotechnological imaginary. Concorde, as *Le Monde* once put it, "was created largely to serve the prestige of France. [It was] the expression of political will, founded on a certain idea of national grandeur" (Harriss 2001). Its development is probably better understood in relation to something like the Moon landings than in relation to civil aviation in isolation.

2. A parallel argument could be made about the industry's unusual commitment to interrogating its failures. Recursive practice is expensive. Many organizations struggle to effectively learn from disasters, not least because ambiguities and interpretive

flexibilities allow them to construct casual interpretations that favor their interests. For example, useful insights are often overlooked if they imply expensive redesigns or costly liabilities (see, e.g., Downer 2014; March et al. 1991; March and Olsen 1988; Hamblin 2012). The civil aviation sphere is far from immune to this trend (see, e.g., Perrow 1983). But with a few notable exceptions (which will be discussed elsewhere in this book), it rarely avoids grasping the nettle where design weaknesses are concerned. Its willingness to do this is interesting for the same reason that its commitment to design stability is interesting, therefore, but exploring the latter offers generalizable insights that explain both behaviors.

3. It has long been understood, for instance, that relocating engines to overwing nacelles would improve both the efficiency and acoustic properties of jetliners (Berguin et al. 2018). In a similar vein, "double-bubble" or blended "wing-body" designs would theoretically allow jetliners to carry many more passengers at a significantly reduced cost. Neither this, nor overawing nacelles, are new or untested concepts. Experts have experimented with prototype wing-body designs since the 1940s, for example, and the aforementioned B-2 Spirit bomber testifies to the viability of both innovations in large airframes.

4. It should be noted that scholars and industry observers sometimes argue that manufacturers were reluctant to implement advanced composites because the materials' higher manufacturing and repair costs would render airframes uncompetitive (e.g., Lenorovitz, 1991; Rogers 1996, vii; AWST 1990; Tenney et al. 2009, 2–3; Slayton and Spinardi 2016, 51). This argument is less than satisfying, however, for a range of reasons. First, it doesn't explain the industry's eagerness to adopt the new materials piecemeal over successive generations of airframes. (If the 787's composite airframe was a poor cost proposition, then presumably the B777's composite tail should have been less optimal than a traditional tail for the same reasons, and so on.) Second, it is far from clear that composites increased maintenance and production costs. Abandoning aluminum undoubtedly disrupted established production regimes and maintenance practices, but the financial implications of this were highly contested within the industry. For instance, many experts at the time argued that higher individual maintenance costs were more than offset by a reduced frequency of maintenance, and higher raw material costs were offset by reduced labor costs (e.g., Jones 1999, 48; Raman, Graser, and Younossi 2003, viii; Barrie 2003; Mecham, 2003, 2005). And although the 787 program was indeed beset by manufacturing difficulties and delays, the larger part of these issues arose less from the composites than from software and electrical issues, together with Boeing's expanded outsourcing (Associated Press 2009; Waltz 2006; Marsh 2009). Third, arguments about increased maintenance and production costs rarely frame those costs against the substantial fuel economies that composites promised. Finally, the argument that airframers avoided composites because the materials were uncompetitive is contradicted by the parallel argument—often advanced by the same observers (e.g., Slayton and Spinardi 2016, 51)—that Boeing's decision to embrace them in the 787 was prompted by a perceived need to compete

more aggressively with Airbus. Companies rarely try to compete more aggressively by introducing less competitive products. All these arguments are more complex than this note can reasonably accommodate, and none are conclusive, but collectively they paint a compelling picture.

5. The regulatory implications of outsourcing were evident in 2013, for instance, when the FAA grounded all B787s after two significant battery fires. The batteries were built by Yuasa, a Japanese manufacturer subcontracted by Thales, a French manufacturer that designed and built a specific electrical system. The NTSB report on the incident highlighted inherent deficiencies in this chain of oversight, wherein the FAA oversaw Boeing, which oversaw Thales, which oversaw Yuasa (NTSB 2013; Bonnín Roca et al. 2017).

6. "Tacit knowledge"—a term coined by Michael Polanyi (1958)—is used widely in the STS literature (e.g., Collins 1982, 2010; MacKenzie and Spinardi 1996). It refers to the information, skill, and experience that are vital to a task but get marginalized, ignored, or obscured by formal accounts because they are uncodified, uncodifiable, or both.

7. Anonymous personal communication, May 19, 2009.

8. Note that few of the factors that keep aviation regulators from being able to police jetliner designs are specific to the aviation industry. Regulators in other spheres are in a similar position regarding their expertise and staffing, and most have interests that are aligned, to some extent, with the industries they regulate. The *Report of the Deepwater Commission*, for instance, made a point of emphasizing regulators' failure to curb the industry's safety compromises, arguing that it was unable to exercise the "autonomy needed to overcome the powerful commercial interests that . . . opposed more stringent safety regulation" (NCBP 2011, 67).

9. It is worth noting that Concorde, again, offers tangible evidence of regulators' powerlessness to enforce stability in defiance of powerful structural interests. Upon welcoming the airplane's inaugural US flight, Undersecretary of Transportation John Barnum declared that it had "met with ease all safety and technical requirements" (Simons 2012, loc. 4717–4722). We could call this a diplomatic truth, but it is probably more accurately described as a lie. US regulators had very serious concerns about the airframe's novel design, which violated a range of several seemingly unambiguous certification rules. (Its engines were not separated on the wing, for example—a factor that became significant in its fatal accident—and it was unable to meet regulations governing minimum fuel reserves [Donin 1976, 54–7]). Concorde had met its US safety and technical requirements, true, but only because those requirements had been warped to accommodate the new design, the regulators' misgivings having been overruled after intervention from the highest levels of government (Simons 2012, loc. 3752–3764, 3967–3976, 3777–3779). The UK and France were both deeply invested in their jetliner's success, and to deny approval would have been to instigate a fierce diplomatic row, and potentially a costly trade war (Simons 2012; Harriss

2001). More broadly, the approval reflected a tangible sense that supersonic flight was the future, and authorities needed to bow to the inevitable lest the US be "left behind" (Simons 2012, loc. 3379–3380).

10. The most notable of these incidents were the losses of American Airlines Flight 191 in May 1979 and United Airlines Flight 232 (the triple hydraulic failure discussed in chapter 6) in July 1989.

CHAPTER 11

1. The MAX was fitted with two AOA indicators, but MCAS took information from only one of them. The airplane could be fitted with an alert to indicate when the two indicators disagreed, but this was an optional extra, for which Boeing charged more.

2. Despite Boeing being the dominant partner in the merger, former McDonnell executives wound up taking leading roles in Boeing's management, giving rise to a longstanding joke about McDonnell effectively buying Boeing with Boeing's money (See, e.g., Useem 2019).

3. The NTSB's recommendation came in response to a DC-8 that overran a runway in Alaska in 1970. The forces involved in the accident were easily survivable, but aircraft had caught fire, and 47 of the 229 people on board died. The postcrash investigation found cyanide in the blood of the bodies recovered and determined that most who died were killed not by the fire directly, but by toxic smoke, which could have been avoidable. The finding led it to recommend that the regulators begin to explore the fire toxicity of cabin interiors (NTSB 1972). In the wake of the NTSB recommendation the FAA established a research effort led by the Civil AeroMedical Institute (CAMI) in Oklahoma City. It concluded that among the many materials that make up a jetliner cabin—everything from seat furnishings to wire claddings— were substances that produced toxic fumes such as carbon monoxide and hydrogen cyanide when heated or burned.

4. Cabin fires are complicated phenomena. Burning conditions can dramatically alter smoke composition and toxicity, for example, even when the same materials are involved (Chatuverdi 2010, 2). So it was that in the decades-long contest over fire-toxicity regulations, airframers questioned the FAA's scientific justifications on a wide number of fronts. The kinds of uncertainties that they raised are well illustrated, however, by a series of simulated cabin-fire tests that the FAA conducted in 1978, with old Lockheed C-133 fuselage. The tests all resulted in "flashovers": explosive conflagrations that would almost inevitably have killed passengers long before toxic fumes became an issue. And experts concluded from this that measures to mitigate toxic fumes were pointless, and regulatory efforts should focus instead on preventing fires in the first place. Others, however, questioned the tests' representativeness. They argued that flashovers occur when smoke and fumes are trapped in a poorly ventilated compartment where they can concentrate and heat until they ignite, and the FAA's

test fuselage—an intact structure, fire-hardened for reuse and largely sealed to out-side ventilation—provided an ideal space for this to occur. But in most real aviation fires, they contended, cabins tend to be much better ventilated, partly because the accidents that instigate them often rupture the fuselage, but also, more prosaically, because people open the doors. Because of this, they claimed, flashovers were much less likely than the tests suggested, and toxicity much more important. The accident record spoke for itself, they protested. Fires did not always involve flashovers, and passengers were demonstrably dying with toxic levels of combustion products in their blood. With experts disagreeing on the meaning of its tests, the FAA eventually conceded that more research was required before it could legitimately require the industry to invest in changes, a position that it still held two decades later (Weir 2000, 87–88; Flight Safety Foundation 1994; Chatuverdi and Sanders 1995, 1; Chatuverdi 2010, 2; ETSC 1996).

5. Perrow, ever insightful, comments on this relationship, albeit in passing. The aviation industry is preoccupied with accident prevention over damage mitigation, he notes, because reducing the injury and death rate from accidents "has little or no effect upon [its] economic variables" (Perrow 1999, 163).

CHAPTER 12

1. Elements of the type-certification process seem actively designed to hide uncertainties and disagreements from public view. Take, for instance, the way that it treats "issue papers": documents that codify expert misgivings and points of contention that arise during assessment. If made public, they would offer a window into the indeterminacies and negotiations that underpin key rulings, but the process designates them as "draft" documents and exempts them from disclosure (NTSB 2006, 42–43, 51).

2. Most trace the roots of this milieu at least as far back as the nineteenth century, when a slew of probabilistic tools emerged as both the product and the driver of new efforts to order and manage the economic world. By many accounts, however, formal audit practices really came of age in the early- to mid-twentieth century—metastasizing, rule by rule, metric by metric, into the daunting modern architecture of statutes, practices, and regulations that Power (1997) calls the "audit society." Chroniclers of this process highlight a wide variety of processes and protagonists in telling this story, including but not limited to the Army Corps of Engineers, flood control, Taylorism, operations research, the RAND Corporation, Cold War defense procurement, the financial services industry, and management consultancy. Many locate its primary drivers in the interfaces between organizational structures, construing audit practices as products of efforts to control and communicate (e.g., Hacking 1990; Porter 1995; Lampland 2010; Desrosieres 1998; Verran 2012; Power 1997).

3. Another structural engineer, A. R. Dykes, artfully captured the essence of white-boxing in a 1976 speech to the British Institution of Structural Engineers when he described his discipline as "the art of modeling materials we do not wholly understand

into shapes we cannot precisely analyze as to withstand forces we cannot properly assess, in such a way that the public has no reason to suspect the extent of our ignorance" (Schmidt 2009, 9).

4. A similar sentiment is evident in Burrows's (2010, 560) epitaph on the accident: "Challenger was lost because NASA came to believe its own propaganda." He writes. "The agency's deeply impacted cultural hubris had it that technology—engineering—would always triumph over random disaster if certain rules were followed. The engineers-turned-technocrats could not bring themselves to accept the psychology of machines with abandoning the core principle of their own faith: equations, geometry, and repetition—physical law, precision design, and testing—must defy chaos. No matter that astronauts and cosmonauts had perished in precisely designed and carefully tested machines. Solid engineering could always provide a safety margin, because the engineers believed, there was complete safety in numbers."

5. Anonymous interview, March 29, 2005.

6. Anonymous interview (February 2, 1996), courtesy of Donald MacKenzie.

CHAPTER 13

1. Much the same might be said of Perrow's normal accidents. Indeed the performance of modern jetliners implies that such accidents are extremely rare, even by the standards required of catastrophic technologies.

2. In an absolute sense, it is rare that reactors melt down, drilling platforms explode, financial instruments crash, or errant deterrence networks almost instigate atomic wars, but such disasters and near-misses occur far more often than experts promise and predict (e.g., Ramana 2011; Sagan 1993). Statistically, this record should be damaging to the idea that ultra-high reliabilities are achievable and knowable, but it is not.

3. It is worth restating here that aviation safety ultimately depends on a wide spectrum of further requirements regarding, for instance, pilot training, maintenance operations, and much more. Navigating the epistemology is only one piece of the puzzle, even if it is a vital piece.

4. Such was the conclusion of Parker Jones, a supervisor of nuclear safety at Sandia National Laboratories, who wrote a report on the accident in 1969, which was declassified in 2013 (Pilkington 2013). And even the conclusion of Secretary of Defense Robert McNamara (quoted in Schlosser 2013, 301), who also references a similar incident in Texas.

CHAPTER 14

1. As the disaster unfolded, the US State Department, unlike several of its European counterparts, chose not to evacuate its embassy in Tokyo. It justified this decision by referring to the Nuclear Regulatory Commission's own radiological hazard modeling,

which, it claimed, showed absolutely no threat to the city. What it didn't say is that the hazard model—named RASCAL, and designed to predict small leaks—maxed out at a range far short of the dangers that it was being used to predict, and so it could not have shown a threat to Japan's capital under any circumstances (Lochbaum et al. 2014, 63).

2. In the discourse around Fukushima, this idea has come to be known as "the myth of absolute safety" (e.g., Hirano 2013; Srinivasan 2013; Onishi 2011; Soble 2014; Nöggerathet al. 2011).

3. As in civil aviation, there is now a tension between the practice and the portrayal of reactor safety assessment. In their internal discourse, leading reactor assessment experts speak eloquently about the subjectivities and limitations of failure predictions. But in public contexts, the same experts vigorously promulgate an understanding of those assessments as rule governed and definitive (Apostolakis 1990; 2004; 1988; Wu and Apostolakis 1992, 335; Wellock 2017, 694; Miller 2003, 183–184). See, for instance, the NRC guidelines for external risk communication. (NRC 2004). Intended for internal consumption, the document instructs regulators to avoid any equivocation in their public assertions of reactor safety because professions of uncertainty "reinforce feelings of helplessness and lack of individual control" (NRC 2004, 38). "Avoid making statements such as 'I cannot guarantee . . .' or '[t]here are no guarantees in life'," it advises, because "statements like these contribute to public outrage" (NRC 2004, 38).

4. Chapter 13 suggested that the promise of perfect reliability was a problematic epiphenomenon of the positivist portrayal of ultrahigh reliability management in civil aviation. From an historical perspective, however, it is clear that this promise was the intended purpose of portraying reliability this way (albeit in the civil nuclear sphere).

5. Propellor-driven airliners predate jetliners, of course, but then nonnuclear power stations predate reactors.

6. It is worth noting, in passing, that this distinction—wherein jetliners require ultrahigh reliability because of their numbers, and reactors do so because of their extreme hazards—becomes significant if we consider the implications of externalities such as sabotage. Aviation safety assessment explicitly does not cover sabotage (e.g., bombings or hijackings) (FAA 2002a, 9). And it can afford to do this because sabotaged jetliners are, to a very limited extent, societally tolerable. This is not the case with reactors, however, as even one sabotaged reactor might be catastrophic. And, unlike most failure mechanisms, there is no reason to imagine that the number of sabotage attempts should be a direct function of how many reactors there are in operation. (This is, admittedly, a complex issue, with considerations not easily captured in a note. Reactors are presumably easier to defend against bombs and hijackers than jetliners are, for instance, but then again, hijacked jetliners could pose a threat to reactors.)

7. Like their aviation counterparts, nuclear authorities (and reactor operators) do endeavor to investigate accidents and anomalies for generalizable insights that can be

applied to other reactors (see, e.g., Perin 2005; Schulman 1993, 363). Post-Fukushima, for example, future plants are unlikely to locate all their backup generators underground, where they are vulnerable to flooding. These efforts are severely constrained by the industry's limited service experience and design stability, however, and, especially in the context of catastrophic failure, they often exhibit a spectrum of weaknesses identified by March, Sproull, and Tamuz (1991, 6–7) in an essay on the problems of extrapolating from limited data. (See also Downer 2014, 2015, 2016; Hamblin 2012; Perin 2005; and Rip 1986).

8. The Price-Anderson Act serves this function in the US, but such statutes are common to almost all nuclear states because the insurance industry is unwilling to cover reactors against catastrophic accidents.

9. The industry invariably counts reactors separately when calculating "cumulative safe operational hours," but groups them into plants when calculating "cumulative accidents."

10. A 2010 NRC study reviewed the implications of the revised seismology for US reactors. It found that earthquake risks to some were significantly higher than had been assumed, to the point where potential quakes might exceed some plants' design bases (NRC 2010; Dedman 2011; Lochbaum et al. 2014, 115). The report ranked plants by risk, finding Indian Point Unit 3, thirty-five miles north of Manhattan, to be the most vulnerable.

REFERENCES

Abernathy, W., and Utterback, J. (1978). "Patterns of Industrial Innovation." *Technology Review* 50: 41–47.

Acohido, B. (1991). "Boeing Co. Pushing the 'Envelope'—Could Early Stops Erode 777's Margin of Safety?" *Seattle Times*, April 21. http://community.seattletimes .nwsource.com/archive/?date=19910421&slug=1278761 (accessed December 2, 2015).

Acohido, B. (1996). "Pittsburgh Disaster Adds to 737 Doubts," *Seattle Times*, October 29.

Acton, J., and Hibbs, M. (2012). "Fukushima Could Have Been Prevented." *New York Times*, March 9, 2012. https://www.nytimes.com/2012/03/10/opinion/fukushima -could-have-been-prevented.html (accessed August 8, 2018).

Aerospace Industries Association of America, Inc. (AIAA). (September 1989). "Maintaining a Strong Federal Aviation Administration: The FAA's Important Role in Aircraft Safety and the Development of US Civil Aeronautics." Washington, D.C.: Aerospace Research Center.

Ahmed, H. and Chateauneuf, A. (2011). "How Few Tests Can Demonstrate the Operational Reliability of Products." *Quality Technology & Quantitative Management* 8: 411–428.

Air Accidents Investigation Branch (AAIB). (February 1976). "Turkish Airlines DC-10, TC-JAV. Report on the Accident in the Ermenonville Forest, France, on March 3, 1974." UK Air Accidents Investigation Branch.

Air Accidents Investigation Branch (AAIB). (1989). "Report on the Accident to Concorde 102, G-BOAF over the Tasman Sea, about 140 nm East of Sydney, Australia, on 12 April 1989." Report No: 6/1989. Department of Transport. Her Majesty's Stationery Office: London.

Air Accidents Investigation Branch (AAIB). (1990). "Report on the Accident to Boeing 737-400 G-OBME near Kegworth, Leicestershire, on 8 January 1989." Aircraft Accident Report 4/90. Air Accidents Investigation Branch. Department of Transport. London.

Air Accidents Investigation Branch (AAIB). (1993). "Report on the Accident to British Aircraft Corporation/SNIAS Concorde 102, G-BOAB, over the North Atlantic, on 21 March 1992." Report No: 5/1993. Department of Transport. Her Majesty's Stationery Office: London.

Airbus. (2017). "A Statistical Analysis of Commercial Aviation Accidents 1958/2017." https://www.airbus.com/content/dam/corporate-topics/publications/safety-first/Airbus-Commercial-Aviation-Accidents-1958-2017.pdf (accessed March 1, 2019).

Airline Pilots Association International (ALPA). (March 1999). "Comments on Rules Docket (AGC-200) Docket No. FAA-1998–4815 (52539)."

Airsafe. (2008). "Significant Safety Events for Air Transit." Airsafe.com, February 2, 2008. http://www.airsafe.com/events/airlines/transat.htm (accessed July 1, 2015).

Airsafe. (2010). "Boeing 747 Plane Crashes." Airsafe.com. http://www.airsafe.com/events/models/boeing.htm (accessed April 10, 2010).

Airsafe. (2020). "Plane Crash Rates by Model." Airsafe.com. http://www.airsafe.com/events/models/rate_mod.htm (accessed July 28, 2020).

Air Safety Week. (2001). "Past Achievements in Safety No Grounds for Complacency." *Air Safety Week*, Monday, August 20.

Air Safety Week. (2005). "Hercules Crash in Baghdad Points to Metal Fatigue in C130's Wing Center." *Air Safety Week*, February 21, 2005.

Air Transport Action Group (ATAG). (April, 2014). "Aviation Benefits beyond Borders." http://aviationbenefits.org/downloads/ (accessed September 15, 2015).

Allen, J. (2004). "Joseph P. Allen Interviewed by Jennifer Ross-Nazzal; Washington DC; 18 March 2004." NASA Johnson Space Center Oral History Project, *Oral History 3 Transcript*. https://historycollection.jsc.nasa.gov/JSCHistoryPortal/history/oral_histories/AllenJP/AllenJP_3-18-04.htm (accessed March 10, 2023).

American Nuclear Society. (March 12, 2011). "Japanese Earthquake/Tsunami; Problems with Nuclear Reactors." http://209-20-84-91.slicehost.net/assets/2011/3/13/ANS_Japan_Backgrounder.pdf (accessed June 19, 2013).

Anderson, A. (2011). "Aviation Safety: Evolution of Airplane Interiors." *Boeing Aeromagazine*. https://www.boeing.com/commercial/aeromagazine/articles/2011_q4/pdfs/AERO_2011q4_article2.pdf (accessed May 19, 2018).

Apostolakis, G. E. (1988). "The Interpretation of Probability in Probabilistic Safety Assessments." *Reliability Engineering & System Safety* 23(4): 247–252.

Apostolakis, G. E. (1990). "The Concept of Probability in Safety Assessments of Technological Systems." *Science* 250(4986): 1359–1364.

Apostolakis, G. E. (2004). "How Useful Is Quantitative Risk Assessment?" *Risk Analysis* 24(3): 515–520.

Arrow, K. J. (1962). "The Economic Implications of Learning by Doing." *Review of Economic Studies* 29(3): 155–173.

Arthur, B. (1989). "Competing Technologies, Increasing Returns, and Lock-in by Historical Events." *Economic Journal* 99: 116–131.

Arthur, B. (2009). *The Nature of Technology: What It Is and How It Evolves.* New York: Free Press.

Associated Press. (2009). "History of the Boeing 787." *Seattle Times*, June 23, 2009. https://web.archive.org/web/20130606210919/http://seattletimes.com/html/nation world/2009373399_apusboeing787historyglance.html (accessed April 4, 2019).

Aubury, M. (1992) "Lessons from Aloha." *BASI Journal*, June 1992. http://www.iasa .com.au/folders/Safety_Issues/others/lessonsfromaloha.html (accessed March 10, 2023).

Australian Transport Safety Bureau (ATSB). (2007). "In-flight Upset Event 240 km North-west of Perth, WA Boeing Company 777-200, 9M-MRG 1 August 2005." *Aviation Occurrence Report*—200503722. http://www.atsb.gov.au/publications/investigation _reports/2005/AAIR/aair200503722.aspx (accessed July 7, 2015).

Australian Transport Safety Bureau (ATSB). (June 27, 2013). "Final Investigation Report, AO-2010-089—In-flight Uncontained Engine Failure Overhead Batam Island, Indonesia 4 November 2010 VH-OQA Airbus A380-842." *ATSB Transport Safety Report.* Canberra.

Aviation Safety & Security Digest. (January 2, 2009). "Safety of Composite Structures Being Evaluated after Aircraft Design Already Approved by FAA." *Aviation Safety & Security Digest.*

 Aviation Week and Space Technology. (January 24, 1990). "Composites May Cut Costs." *Aviation Week and Space Technology.*

Bainbridge, L. (1983). "Ironies of Automation." *Automatica* 19(6): 775–777.

Barcott, B. (2009). "The Role of Suburban Sprawl." *New York Times*, January 16.

Barker, I. (2006). "Composite Planes" Letter to the Editor. *New Scientist*, August 30, 2006. https://www.newscientist.com/letter/mg19125673-200-composite-planes/ (accessed August 9, 2016).

Barlay, S. (1990). *The Final Call: Why Airline Disasters Continue to Happen.* New York: Pantheon Books.

Barrie, D. (2003). "Boeing Targets Its 7E7 at Mid-Market Gap." *Aviation Week and Space Technology*, February 3, 2003.

Baxandall, M. (1985). *Patterns of Intention: On the Historical Explanation of Pictures.* New Haven, CT: Yale University Press.

Bazerman, M. H., and Watkins, M.D. (2004). *Predictable Surprises: The Disasters You Should Have Seen Coming and How to Prevent Them.* Boston: Harvard Business School Press.

Bazovsky, I. (1961). *Reliability Theory and Practice.* Hoboken, NJ: Prentice-Hall.

BBC. (2015). "AirAsia Crash: Faulty Part 'Major Factor'." *News Online*, December 1, 2015. Online: https://www.bbc.co.uk/news/world-asia-34972263 (accessed May 5, 2018).

Beatson, J. (1989). "Air Safety: Is America Ready to 'Fly By Wire'?" *Washington Post*, April 2, 1989. https://www.washingtonpost.com/archive/opinions/1989/04/02/air

-safety-is-america-ready-to-fly-by-wire/029882be-28cb-4f56-8fe4-1f48e333eb58/?nore
direct=on&utm_term=.9d8718964e33 (accessed April 4, 2019).

Beck, U. (1992). *Risk Society: Towards A New Modernity*. London: SAGE.

Beck, U. (1999). *World Risk Society*. London: Polity Press.

Benaroya, H. (2018). *Building Habitats on the Moon: Engineering Approaches to Lunar Settlements*. London: Springer.

Berguin, H. S., Renganathan, A., Ahuja, J., Chen, M., Perron, C., Tai, J. and Mavris, D. (2018). *CFD Study of an Over-Wing Nacelle Configuration*. Airbus Technical Report No. 1853/60464. October.

Bier, V., Joosten, J. Glyer, D. Tracey, J., and Welsh, M. (2003). *Effects of Deregulation on Safety: Implications Drawn from the Aviation, Rail, and United Kingdom Nuclear Power Industries*. New York: Springer Science & Business Media.

Bijker, W. Hughes, T., and Pinch, T. (Eds.) (1989). *The Social Construction of Technological Systems: New Directions in the Sociology and History of Technology*. Cambridge, MA: MIT Press.

Bishop, P. G. (1995). "Software Fault Tolerance by Design Diversity." In Lyu, M. (Ed.), *Software Fault Tolerance*, 211–229. Hoboken, NJ: John Wiley & Sons.

Blastland, M., and Spiegelhalter, D. (2013). *The Norm Chronicles: Stories and Numbers about Danger*. London: Profile Books.

Blockley, D. (2012). *Engineering: A Very Short Introduction*. Oxford: Oxford University Press.

Bloor, D. (1976). *Knowledge and Social Imagery*. London: Routledge & Kegan Paul.

Bó, E. D. (2006). "Regulatory Capture: A Review." *Oxford Review of Economic Policy* 22(2): 203–225.

Bohn, R. E. (2005). "From Art to Science in Manufacturing: The Evolution of Technological Knowledge." *Foundations and Trends in Technology, Information and Operations Management* 1(2): 1–82.

Bokulich, F. (2003). "Birdstrikes Remain a Concern for Pilots." *Technology Update*, May 11. http://www.sae.org/aeromag/techupdate_3-00/05.htm. (accessed July 3, 2012).

Bonnín Roca, J., Vaishnav, P., Morgan, M. G., Mendonça, J., and Fuchs, E. (2017). "When Risks Cannot Be Seen: Regulating Uncertainty in Emerging Technologies." *Research Policy* 46(7): 1215–1233.

Boston, C. (2019). "Hedge Funds Resurrect CDO Trade. This Time They Say It Will Work." *Bloomberg*, May 2. https://www.bloomberg.com/news/articles/2019-05-02/hedge-funds-resurrect-cdo-trade-this-time-they-say-it-will-work (accessed September 14, 2021).

Bostrom, N. (2014) *Superintelligence: Paths, Dangers, Strategies*. Oxford: Oxford University Press.

Bowker, G., and Star, S.L. (2000). *Sorting Things Out: Classification and Its Consequences*. Cambridge, MA: MIT Press.

Briddon, A. Ellmore, C., and Marraine, P. (1974). *FAA Historical Fact Book: A Chronology 1926–1971*. Washington, DC: US Government Printing Office.

Brown, D. P. (2009). "All Nippon Airways Wants Passengers to Use the Bathroom before Boarding." *AirlineReporter*, October 13, 2009. https://www.airlinereporter.com /2009/10/all-nippon-airways-wants-passengers-to-use-the-bathroom-before-boarding/ (accessed September 22, 2016).

Bruce, J. T., and Draper, J. (1970). *[The NaderReport on] Crash Safety in General Aviation Aircraft*. Washington, DC: Center for Study of Responsive Law.

Bucciarelli, L. (1994). *Designing Engineers*. Cambridge, MA: MIT Press.

Budgey, R. (1999). "Three-Dimensional Bird Flock Structure and Its Implications for Birdstrike Tolerance in Aircraft." Paper delivered to International Bird Strike Committee, September 1999 (IBSC24/ WP 12).

Budgey, R. (2000). "The Development of a Substitute Artificial Bird By the International Birdstrike Research Group for Use in Aircraft Component Testing." International Bird Strike Committee IBSC25/WP-IE3, Amsterdam: 17–21.

Bugos, G. (1996). *Engineering the F4 Phantom II: Parts into Systems*. Annapolis, MD: Naval Institute Press.

Bureau Enquetes-Accidents (BEA). (June 1980). "Rapport D'Enquete Concernant L'Incident Survenu le 14 Juin 1979 a Washington-Dulles (Etats-Unis) au Concorde No. 9 Immatriculre F-BVFC." Paris: Ministere Des Transports.

Bureau Enquêtes-Accidents (BEA). (January 2002). "Accident on 25 July 2000 at La Patte d'Oie in Gonesse (95) to the Concorde Registered F-BTSC Operated by Air France." Report translation f-sc000725a. Paris: Ministere Des Transports.

Bureau Navigabilite des Moteurs et Equipements (BNME). (1999). "Commentaires NPRM 98-1 9." Bureau Navigabilite Des Moteurs et Equipements, Docket No FAA-98-2815-10 (53165), March 31.

Burnie, S. (2017). "Fukushima Bill." *Asia Times*, March 31, 2017. http://www.atimes .com/article/tepcos-fukushima-expensive-industrial-accident-history/ (accessed July 25, 2017).

Burrows, W. (2010). *This New Ocean: The Story of the First Space Age*. London: Random House.

Butler, R., and Finelli, G. (1993). "The Infeasibility of Experimental Quantification of Life-Critical Software reliability." *IEEE Transactions on Software Engineering* 19(1): 3–12.

Calder, S. (2018). "Airline Safety: 2017 Was Safest Year in History for Passengers around World, Research Shows." *The Independent*, January 1. https://www.independent.co.uk /travel/news-and-advice/air-safety-2017-best-year-safest-airline-passengers-worldwide -to70-civil-aviation-review-a8130796.html (accessed September 21, 2018).

Calma, J. (2020). "Offshore Drilling Has Dug Itself a Deeper Hole." *The Verge*, April 20. https://www.theverge.com/2020/4/20/21228577/offshore-drilling-deepwater-horizon -10-year-anniversary (accessed September 9, 2021).

Cannon, R., and Zimmermann, R. (1985). "Seat Experiments for the Full-Scale Transport Aircraft Controlled Impact Demonstration." DOT/FAA/CT-84/10 March. Atlantic City, NJ: Federal Aviation Administration Technical Center.

Capaccio, T. (1997). "The B-2's Stealthy Skins Need Tender, Lengthy Care." *Defense Week*, May 27, 1997: 1.

Capaccio, T. (2019). "Stagnant F-35 Reliability Means Fewer Available Jets: Pentagon." *Bloomberg Government*. https://about.bgov.com/news/stagnant-f-35-reliability-means-fewer-available-jets-pentagon/ (accessed March 14, 2019).

Chaplin, J. C. (2011). "Safety Regulation—The First 100 Years." *Journal of Aeronautical History* 3: 75–96.

Chaturvedi, A. K. (2010). "Aviation Combustion Toxicology: An Overview." *Journal of Analytical Toxicology* 34: 1–16.

Chatuverdi, A., and Sanders, D. (February 1995). Aircraft Fires, Smoke Toxicity, and Survival: An Overview. CAMI Report No. 95/8. Oklahoma City: Civil Aeromedical Institute.

Chokshi, N. (2020). "Airlines May Learn to Love the Boeing 737 Max Again." *New York Times*, July 15. https://www.nytimes.com/2020/07/15/business/boeing-737-max-return.html (accessed July 7, 2020).

Civil Aviation Authority (CAA). (2013). "Global Fatal Accident Review: 2002 to 2011." CAP 1036. London: Stationery Office.

Clarke, J. D. (2017 [1972]). *Ignition! An Informal History of Rocket Propellants*. New Brunswick, NJ: Rutgers University Press.

Clarke, L. (1989). *Acceptable Risk? Making Decisions in a Toxic Environment*. Berkeley: University of California Press.

Clarke, L. (2005). *Worst Cases: Terror and Catastrophe in the Popular Imagination*. Chicago: University of Chicago Press.

Clarke, L., and Perrow C. (1996). "Prosaic Organizational Failure." *American Behavioral Scientist* 39(8): 1040–1056.

Cleary, E. C., Dolbeer, R. A., and Wright, S. E. (2006). "Wildlife Strikes to Civil Aircraft in the United States, 1990–2005." Washington, DC: US Department of Transportation, Federal Aviation Administration, Office of Airport Safety and Standards.

Cobb, R., and Primo, D. (2003). *The Plane Truth: Airline Crashes, the Media, and Transportation Policy*. Washington, DC: Brookings Institution Press.

Cohen, M., March, J., and Olsen, J. (1972). "A Garbage Can Model of Organizational Choice." *Administrative Science Quarterly* 17 (1): 1–25.

Collins, H. (1981). "The Place of the 'Core-Set' in Modern Science: Social Contingency with Methodological Propriety in Science." *History of Science* (19): 6–19.

Collins, H. (1982). "Tacit Knowledge and Scientific Networks." In Barnes B. and Edge D. (Eds.), *Science in Context: Readings in the Sociology of Science*, 44–64. Milton Keynes, UK: Open University Press.

Collins, H. (1985). *Changing Order*. London: SAGE.

Collins, H. (1988). "Public Experiments and Displays of Virtuosity: The Core-Set Revisited." *Social Studies of Science* 18(4): 725–748.

Collins, H. (2001). "Tacit Knowledge and the Q of Sapphire." *Social Studies of Science* 31(1): 71–85.

Collins, H. (2010). *Tacit and Explicit Knowledge*. Chicago, IL: University of Chicago Press.

Collins, H., and Evans, R. (2002). "The Third Wave of Science Studies: Studies of Expertise and Experience." *Social Studies of Science* 32(2): 235–296.

Collins, H., Evans, R., and Weinel, M. (2017). "STS as Science or Politics?" *Social Studies of Science* 47(4): 587–592.

Collins, H., and Pinch, T. (1993). *The Golem: What Everyone Should Know about Science*. Cambridge: Cambridge University Press.

Collins H., and Pinch, T. (1998). *The Golem at Large: What You Should Know about Technology*. Cambridge: Cambridge University Press.

Constant, E. (1980). *The Origins of the Turbojet Revolution*. Baltimore: Johns Hopkins University Press.

Constant, E. (1999). "Reliable Knowledge and Unreliable Stuff: On the Practical Role of Unreliable Beliefs." *Technology and Culture* 40(2): 324–357.

Coppola, A. (1984). "Reliability Engineering of Electronic Equipment: An Historical Perspective." *IEEE Transactions on Reliability* R-33(1): 29–35.

Cowen, H. (1972). "Fairey Fly-by-Wire." *Flight International*, August 10, 1972: 193.

Crane, D. (1997). *Dictionary of Aeronautical Terms*, 3rd ed. Newcastle, WA: Aviation Supplies & Academics.

Croft, J. (2009). "NTSB, FAA Investigate High Altitude Bird Strikes Near Phoenix." *FLightGlobal*, November 6, 2009. https://www.flightglobal.com/news/articles/ntsb-faa-investigate-high-altitude-bird-strikes-nea-334523/ (accessed June 18, 2018).

Cronrath, E. M. (2017). *The Airline Profit Cycle: A System Analysis of Airline Industry Dynamics*. London: Routledge.

Cushman, J. (1989). "U.S. Investigators Fault Aloha Line in Fatal Accident." *New York Times*, May 24.

Dana, D., and Koniak, S. (1999). "Bargaining in the Shadow of Democracy." *University of Pennsylvania Law Review* 148(2): 473–559.

David, P. A. (1985). "Clio and the Economics of QWERTY." *American Economic Review Papers and Proceedings* 75: 332–337.

Davies, R. (1986). "The Effect of the Accident at Chernobyl upon the Nuclear Community." *Science, Technology, & Human Values* 11(4): 59–63.

Davies, R., and Birtles, P. J. (1999). *Comet: The World's First Jet Airliner*. Richmond, VA: Paladwr Press.

de Briganti, G. (2016). "Navair Sees F-35 Requiring up to 50 Maintenance Hours per Flight Hour." *Defense-Aerospace*, December 5, 2016. http://www.defense-aerospace .com/articles-view/feature/5/179243/navair-projects-f_35-to-need-50-maintenance -hours-per-flight-hour.html (accessed March 10, 2019).

Dedman, B. (2011). "What Are the Odds? US Nuke Plants Ranked by Quake Risk." NBC News. http://www.nbcnews.com/id/42103936/ns/world_news-asiapacific/ (accessed May 17, 2017).

Dekker, S. (2011). *Drift into failure. From hunting broken components to understanding complex systems.* Aldershot, UK: Ashgate: Ashgate.

Desrosieres, A. (1998). *The Politics of Large Numbers: A History of Statistical Reasoning.* Cambridge, MA: Harvard University Press.

DeVault, T. L., Blackwell, B. F., Seamans, T. W., Lima, S. L. and Fernández-Juricic, E. (2014). "Effects of Vehicle Speed on Flight Initiation by Turkey Vultures: Implications for Bird-Vehicle Collisions." *PLoS One* 9(2): e87944.

DeVault, T. L., Blackwell, B. F., Seamans, T. W., Lima, S. L., and Fernández-Juricic, E. (2015). "Speed Kills: Ineffective Avian Escape Responses to Oncoming Vehicles." *Proceedings. Biological Sciences/The Royal Society* 282(1801): 20142188.

Dolbeer, R. (2007). "Bird Damage to Turbofan and Turbojet Engines in Relation to Phase of Flight—Why Speed Matters." *Bird and Aviation* 27(2): 1–7.

Dolbeer, R., and Eschenfelder, P. (2003). "Amplified Bird-Strike Risks Related to Population Increases of Large Birds in North America." *Proceedings of the 26th International Bird Strike Committee meeting (Volume 1).* Warsaw, Poland: 49–67.

Donin, R. (1976). "Safety Regulation of the Concorde Supersonic Transport: Realistic Confinement of the National Environmental Policy Act." *Transportation Law Journal* 8: 47–69.

Douglas, M., and Wildavsky, A. B. (1982). *Risk and Culture: An Essay on the Selection of Technical and Environmental Dangers.* Berkeley: University of California Press.

Downer, J. (2007). "When the Chick Hits the Fan: Representativeness and Reproducibility in Technological Testing." *Social Studies of Science* 37(1): 7–26.

Downer, J. (2009a). "Watching the Watchmaker: On Regulating the Social in Lieu of the Technical." *LSE CARR Discussion Paper 54*, June. http://eprints.lse.ac.uk/36538/ (accessed March 10, 2023).

Downer, J. (2009b). "When Failure *Is* an Option: Redundancy, Reliability, and Risk." *LSE CARR Discussion Paper 53*, May. https://eprints.lse.ac.uk/36537/1/Disspaper53 .pdf (accessed March 10, 2023).

Downer, J. (2010). "Trust and Technology: The Social Foundations of Aviation Regulation." *British Journal of Sociology* 61(1): 87–110.

Downer, J. (2011a). "On Audits and Airplanes: Redundancy and Reliability-Assessment in High Technologies." *Accounting, Organizations and Society* 36(4): 269–283.

Downer, J. (2011b). "'737-Cabriolet': The Limits of Knowledge and the Sociology of Inevitable Failure." *American Journal of Sociology* 117(3): 725–762.

Downer, J. (2014). "Disowning Fukushima: Managing the Credibility of Nuclear Reliability Assessment in the Wake of Disaster." *Regulation & Governance* 8: 287–309.

Downer, J. (2015). "The Unknowable Ceilings of Safety: Three Ways That Nuclear Accidents Escape the Calculus of Risk Assessments." In Taebi, B. and Roeser, S. (Eds.), *The Ethics of Nuclear Energy: Risk, Justice and Democracy in the Post-Fukushima Era*, 35–52. Cambridge: Cambridge University Press.

Downer, J. (2016). "Resilience in Retrospect: Interpreting Fukushima's Disappearing Consequences." In Herwig, A. and Simoncini, M. (Eds.), *Law and the Management of Disasters: The Challenge of Resilience*, 42–60. London: Routledge.

Downer, J. (2020). "On Ignorance and Apocalypse: A Brief Introduction to 'Epistemic Accidents.'" In LeCoze, J. C. (Ed.) *New Perspectives on Safety Research*. Boca Raton, FL: CDC Press.

Downer, J., and Ramana, M.V. (2021). "Empires Built on Sand: On Reactor Safety Assessment and Regulation." *Regulation & Governance* 15(1): 1304–1325.

Drier, M. (2020). "Cormorants Damaging Fish Populations." *Huron Daily Tribune*, February 29. https://www.michigansthumb.com/news/article/Cormorants-damaging-fish-populations-15092892.php (accessed August 4, 2020).

Duffy, C. (2020). "Boeing Has Uncovered Another Potential Design Flaw with the 737 Max." *CNN Business*, January 6. https://edition.cnn.com/2020/01/05/business/boeing-737-max-wiring-issue/index.html (accessed July 7, 2020).

Dumas, L. (1999). *Lethal Arrogance: Human Fallibility and Dangerous Technologies*. New York: St. Martin's Press.

Dvorak, P. (2012a). "Fukushima Daiichi's Achilles Heel: Unit 4's Spent Fuel?" *Wall Street Journal*, April 17. http://blogs.wsj.com/japanrealtime/2012/04/17/fukushima-daiichis-achilles-heel-unit-4s-spent-fuel (accessed September 15, 2013).

Dvorak, P. (2012b). "Fukushima Daiichi's Unit 4 Spent-Fuel Pool: Safe or Not?" *Wall Street Journal* (Japan), April 17. http://blogs.wsj.com/japanrealtime/2012/05/21/fukushima-daiichis-unit-4-spent-fuel-pool-safe-or-not/ (accessed August 5, 2013).

Eckhard, D., and Lee, L. (1985). "A Theoretical Basis of Multiversion Software Subject to Coincident Errors." *IEEE Transactions on Software Engineering* 11: 1511–1517.

Edge, C. E., and Degrieck, J. (1999). "Derivation of a Dummy Bird for Analysis and Test of Airframe Structures." *1999 Bird Strike Committee-USA/Canada, First Joint Annual Meeting, Vancouver, BC*. Paper 14. http://digitalcommons.unl.edu/birdstrike1999/14 (accessed March 10, 2023).

Edmondson, A. C. (2011). "Strategies for Learning from Failure." *Harvard Business Review*, April 2011. http://www.preemptivetesting.com/resources/Strategies%20for%20Learning%20from%20Failure.pdf (accessed April 1, 2019).

Endres, G. (1998). *McDonnell Douglas DC-10*. St. Paul, MN: MBI Publishing Company.

Erikson, K. (1991). "Radiation's Lingering Dread." *Bulletin of the Atomic Scientists* 47(2): 34–39.

Eschenfelder, P. (2000). "Jet Engine Certification Standards." Paper delivered to International Bird Strike Committee, April 17–21 (IBSC25/WP-IE1).

Eschenfelder, P. (2001). "Wildlife Hazards to Aviation." Paper delivered to ICAO/ACI Airports Conference. Miami, April 24.

Espeland, W., and Stevens, M. (2008). "A Sociology of Quantification." *European Journal of Sociology* 49: 401–436.

European Transport Safety Council (ETSC). (December 1996). "Increasing the Survival Rate in Accident Conditions: Impact Protection, Fire Survivability and Evacuation." European Transport Safety Council. Brussels.

Ezrahi, Y. (2008). "Technology and the Civil Epistemology of Democracy." *Inquiry* 35(3–4): 363–376.

Ezrahi, Y. (2012). *Imagined Democracies: Necessary Political Fictions*. New York: Cambridge University Press.

Fackler, M. (2012). "Nuclear Crisis Set off Fears over Tokyo, Report Says." In *New York Times*, February 27, 2012. http://www.nytimes.com/2012/02/28/world/asia/japan-considered-tokyo-evacuation-during-the-nuclear-crisis-report-says.html?_r=1 (accessed February 8, 2012).

Faith, N. (1996). *Black Box*. London: Boxtree.

Fallows, J. (1985). "The American Army and the M-16 Rifle." In MacKenzie, D. and Wajcman, J. (Eds.), *The Social Shaping of Technology*, 2nd ed., 382–394. Buckingham, UK: Open University Press.

Favre, C. (1996). "Fly-By-Wire for Commercial Aircraft: The Airbus Experience." In Tischler, M. (Ed.), *Advances in Aircraft Flight Control*. UK: Taylor & Francis.

Fanfalone, M. (2003). *Testimony to the House Aviation Subcommittee: On FAA Reauthorization*. March 27. http://www.findarticles.com/p/articles/mi_m0UBT/is_17_17/ai _100769720#continue.

Federal Register. (1998). "Proposed Rules." Vol. 63, No. 238. December 11: 68641–68644.

Feeler, R. A. (1991). "The Mechanic and Metal Fatigue." *Aviation Mechanics Bulletin*. Flight Safety Foundation. March/April: 1–2.

Feris, M. A. (2003). "It's Murder on the Runways at SA's Airports." *Cape Argus*, September 1.

Feyerabend, P. (1975). *Against Method: Outline of an Anarchistic Theory of Knowledge*. London: Redwood Burn.

Feyerabend, P. (1995). *Killing Time: The Autobiography of Paul Feyerabend*. Chicago: Chicago University Press.

Fiorino, F. (2009). "Survival Is the Rule." *New York Times*, January 16.

Fleck, J. (1994). "Learning by Trying: The Implementation of Configurational Technology." *Research Policy* 23: 637–652.

Flight Safety Foundation. (1994). "Getting out Alive—Would Smoke Hoods Save Airline Passengers or Put Them at Risk?" *Cabin Crew Safety* 29(1): 1–24.

Flight Safety Foundation. (2018). "Fatal Accidents per Year 1946–2017." *Aviation Safety Network*. https://aviation-safety.net/statistics/ (accessed October 10, 2018).

Fraher, A. (2014). *The Next Crash: How Short-Term Profit-Seeking Trumps Airline Safety.* New York: Cornell University Press.

Francis, John (1993) *The Politics of Regulation: A Comparative Perspective.* Oxford: Blackwell.

Freeman, C., and Perez, C. (1988). "Structural Crises of Adjustment, Business Cycles, and Investment Behavior." In Dosi, G., Freeman, C., Nelson, R., Silverberg, G. and Soete, L. (Eds.), *Technical Change and Economic Theory*, 38–66. London: Francis Pinter.

Froud, J., Sukhdev, J., Leaver, A., and Williams, K. (2006). *Financialization and Strategy: Narratives and Numbers.* London: Routledge.

Fukushima Nuclear Accident Independent Investigation Commission (NAIIC). (2012). "The Official Report of the Fukushima Nuclear Accident Independent Investigation Commission (Executive Summary)." Tokyo: National Diet of Japan. http://www.nirs .org/fukushima/naiic_report.pdf (accessed July 16, 2013).

Fuller, J. (1976). *We Almost Lost Detroit.* New York: Ballantine Books.

Galison, P. (1987). *How Experiments End.* Chicago: University of Chicago Press.

Galison, P. (2000). "An Accident of History." In Galison, P. and Roland, A. (Eds.), *Atmospheric Flight in the Twentieth Century*, 3–43. Boston, Kluwer.

Gapper, J. (2014). "The Price of Innovation Is Too High in Aerospace." *Financial Times,* July 16. http://www.ft.com/intl/cms/s/0/7ab776a2-0ce0-11e4-bf1e-00144feabdc0.html (accessed August 26, 2015).

Garrison, P. (2005). "When Airplanes Feel Fatigued." *Flying,* September. https://www .flyingmag.com/when-airplanes-feel-fatigued/ (accessed March 10, 2023).

Gates, D. (2007). "Fired Engineer Calls 787's Plastic Fuselage Unsafe." *Seattle Times*, September 18. http://old.seattletimes.com/html/boeingaerospace/2003889663_boeing180 .html (accessed April 4, 2019).

Gelles, D. (2019). "Boeing Can't Fly Its 737 Max, but It's Ready to Sell Its Safety." *New York Times,* December 24.

Gephart, R., Van Maanen, J., and Oberlechner, T. (2009) "Organizations and Risk in Late Modernity." *Organization Studies* 30(2/3): 141–155.

Gerard, D., and Lave, L. B. (2005). "Implementing Technology-Forcing Policies: The 1970 Clean Air Act Amendments and the Introduction of Advanced Automotive Emissions Controls in the United States." *Technological Forecasting and Social Change* 72(7): 761–778.

Gilligan, A. (2016). "Fukushima: Tokyo Was on the Brink of Nuclear Catastrophe, Admits Former Prime Minister." *The Telegraph*, March 4, 2016. https://www.telegraph .co.uk/news/worldnews/asia/japan/12184114/Fukushima-Tokyo-was-on-the-brink-of -nuclear-catastrophe-admits-former-prime-minister.html (accessed August 8, 2018).

Gillman, P. (January 1977). "Supersonic Bust: The Story of the Concorde." *Atlantic Monthly* 239(1): 72–81.

Goodspeed, P. (2012). "Japan Feared Post-Tsunami 'Devils Chain Reaction.'" *National Post,* February 28. http://fullcomment.nationalpost.com/2012/02/28/peter-goodspeed

-japanese-officials-feared-post-tsunami-devils-chain-reaction-would-destroy-tokyo
-report-finds/ (accessed July 9, 2013).

Gordon, J. E. (2018 [1991]). *Structures: Or Why Things Don't Fall Down*. London: Penguin Books Ltd.

Gorton, G. (2012). *Misunderstanding Financial Crises: Why We Don't See Them Coming*. Oxford: Oxford University Press.

Graeber, D. (2015). *The Utopia of Rules: On Technology, Stupidity, and the Secret Joys of Bureaucracy*. London: Melville House.

Graffeo, E. (2021) "Warren Buffett Warned 18 Years Ago about the Financial Instruments that Triggered the Archegos Implosion." *Business Insider*. April 2, 2021. https://markets.businessinsider.com/news/stocks/warren-buffett-archegos-implosion-warning-derivatives-total-return-swaps-lethal-2021-4-1030272114 (accessed March 12, 2023).

Gritta, R., Adrangi, B., and Davalos, S. (2006). "A Review of the History of Air Carrier Bankruptcy Forecasting and the Application of Various Models to the U.S. Airline Industry: 1980–2005." *Credit and Financial Management Review* XII(3): 11–30.

Gross, P., and Levitt, N. (1994). *Higher Superstition: The Academic Left and Its Quarrels with Science*. Baltimore: Johns Hopkins University Press.

Guston, B. (1995). *The Osprey Encyclopedia of Russian Aircraft 1875–1995*. London: Osprey Publishing.

Hacking, I. (1990). *The Taming of Chance*. Cambridge: Cambridge University Press.

Hallion, R. (2004). *Taking Flight: Inventing the Aerial Age, from Antiquity through the First World War*. Oxford: Oxford University Press.

Hamblin, J. (2007). "'A Dispassionate and Objective Effort:' Negotiating the First Study on the Biological Effects of Atomic Radiation." *Journal of the History of Biology* 40(1): 147–177.

Hamblin, J. (2012) "Fukushima and the Motifs of Nuclear History." *Environmental History* 17: 285–299.

Hansen, M., McAndrews, C., and Berkeley, E. (2008). "History of Aviation Safety Oversight in the United States." *Final Report to Federal Aviation Administration*. Washington, DC: DOT/FAA/AR-08/39.

Harrington, A., and Downer, J. (2019). "*Homo Atomicus*: An Actor Worth Psychologizing? The Problems of Applying Behavioral Economics to Nuclear Strategy." In Knopf, J., and Harrington, A. (Eds.), *Behavioral Economics and Nuclear Weapons*, 187–202. Athens: University of Georgia Press.

Harriss, J. (2001). "The Concorde Redemption: Can the Superplane Make a Comeback?" *Air & Space Magazine*. https://www.airspacemag.com/flight-today/the-concorde-redemption-2394800/?all (accessed February 2, 2019).

Harvey, F. (2011). "Nuclear Is the Safest Form of Power, Says Top UK Scientist." *The Guardian*, March 30. http://www.guardian.co.uk/environment/2011/mar/29/nuclear-power-safe-sir-david-king (accessed June 19, 2013).

Haynes, A. (1991). "The crash of United Flight 232." *Paper presented at NASA Ames Research Center, Dryden Flight Research Facility*, Edwards, California, 24 May.

Hayward, B. and Lowe, A. (2004). "Safety Investigation: Systemic Occurrence Analysis Methods." In Goeters, K. M. (Ed.), *Aviation Psychology: Practice and Research*. United Kingdom: Ashgate: 363–380.

Heimer, C. (1980). "Substitutes for Experience-based Information: The Case of Offshore Oil Insurance in the North Sea." Discussion Paper no. 1181. Bergen, Norway: Institute of Industrial Economics.

Henke, C. (2000). "Making a Place for Science: The Field Trial." *Social Studies of Science* 30(4): 483–511.

Higgins, M. (2010). "Babies on Airlines: Safety Seats Are Safer than a Lap." *New York Times*, November 23.

Hilgartner, S. (2000). *Science on Stage: Expert Advice as Public Drama*. Palo Alto, CA: Stanford University Press.

Hilgartner, S. (2007). "Overflow and Containment in the Aftermath of Disaster." *Social Studies of Science* 37(1): 153–158.

Hilkevitch, J. (2004). "Goose Got the Blame but It Was Rare Bird Plane Hit." *Chicago Tribune*, September 18.

Hirano, S. (2013). "Nuclear Damage Compensation in Japan: Multiple Nuclear Meltdowns and a Myth of Absolute Safety." 総合政策研究, 21: 1–35.

Hiserodt, E. (2011). "Fukushima: Just How Dangerous Is Radiation?" *The New American*, April 27. https://thenewamerican.com/print/fukushima-just-how-dangerous-is-radiation/ (accessed March 10, 2023).

Hogge, L. (2012). *Effective Measurement of Reliability of Repairable USAF Systems*. PhD thesis. AFIT/GSE/12-S02DL. Wright-Patterson AFB, OH: Air Force Institute of Technology.

Holanda, R. (2009). *A History of Aviation Safety Featuring the US Airline System*. Bloomington, IN: Authorhouse.

Hollnagel, E. (2006). "Resilience—the Challenge of the Unstable." In Hollnagel, E., Woods, D. and Leveson, N. (Eds.), *Resilience Engineering: Concepts and Precepts*, 9–17. Aldershot, UK: Ashgate.

Hollnagel, E., Woods, D., and Leveson, N. (Eds.). (2006). *Resilience Engineering: Concepts and Precepts*. Aldershot, UK: Ashgate.

Honolulu Advertiser. (2001). "Engineer Fears Repeat of 1988 Aloha Jet Accident." January 18. http://the.honoluluadvertiser.com/2001/Jan/18/118localnews1.html (accessed March 10, 2023).

Hopkins, A. (1999). "The Limits of Normal Accident Theory." *Safety Science* 32(2): 93–102.

Hopkins, A. (2009). *Failure to Learn: The BP Texas City Refinery Disaster*. North Ryde: CCH Australia.

Hopkins, A. (2010). "Why BP Ignored Close Calls at Texas City." *Risk & Regulation. Special Issue on Close Calls, Near Misses and Early Warnings,* July, 4–5.

Hopwood, A. G. and Miller, P. (Eds.). (1994). *Accounting as Social and Institutional Practice.* Cambridge: Cambridge University Press.

Horgan, J. (2016). "Was Philosopher Paul Feyerabend Really Science's 'Worst Enemy'?" *Scientific American,* October 24. https://blogs.scientificamerican.com/cross-check/was -philosopher-paul-feyerabend-really-science-s-worst-enemy/ (accessed August 8, 2017).

Huber, M. (2004). "Beached Starship." *Smithsonian Air & Space Magazine,* September. http://www.airspacemag.com/military-aviation/beached-starship-5429731/ (accessed September 7, 2016).

Hughes, R. (1987). "A New Approach to Common Cause Failure." *Reliability Engineering* 17: 2111–2136.

Hutter, B. M. (2001). *Regulation and Risk: Occupational Health and Safety on the Railways.* Oxford: Oxford University Press.

Institution of Electrical Engineers (IEE). (2005). *Nuclear Reactor Types.* London. http://large.stanford.edu/courses/2013/ph241/kallman1/docs/nuclear_reactors.pdf (accessed August 21, 2016).

Insurance Information Institute. (2018). "Aviation." https://www.iii.org/fact-statistic /facts-statistics-aviation-and-drones?table_sort_735941=9 (accessed September 21, 2018).

International Air Transport Association (IATA). (2013). "Products and Services Release: The Airline Industry Story for 2012." Press Release, July 16. http://www.iata.org/press room/pr/Pages/2013-07-16-01.aspx (accessed December 20, 2017).

International Civil Aviation Organization (ICAO). (1984). *Accident Prevention Manual.* January 1. Document no. 9422.

Isidore, C. (2019). "Boeing Is about to Reveal Just How Much the 737 Max Crisis Hurt Its Business." *CNN Business.* https://edition.cnn.com/2019/04/08/business /boeing-737-max-deliveries/index.html (accessed April 8, 2019).

Isidore, C. (2020a). "Boeing Says It Found Debris in Fuel Tanks of Parked 737 Max Jets" *CNN Business,* February 19. https://edition.cnn.com/2020/02/19/business/boeing -737-max-fuel-tank-debris/index.html (accessed July 7, 2020).

Isidore, C. (2020b). "The Cost of the Boeing 737 Max Crisis: $18.7 Billion and Counting." *CNN Business,* March 10. https://edition.cnn.com/2020/03/10/business /boeing-737-max-cost/index.html (accessed July 7, 2020).

Jansen, B. (2014). "NTSB Chief Urges Child-Safety Seats on Planes." *USA Today,* April 21. http://www.usatoday.com/story/news/nation/2014/04/21/ntsb-hersman -national-press-club-plane-crashes-rail-safety-buses-cars/7961795/ (accessed August 6, 2015).

Jasanoff, S. (1986). *Risk Management and Political Culture.* New York: Russell Sage Foundation.

Jasanoff, S. (1987). "Contested Boundaries in Policy-Relevant Science." *Social Studies of Science* 17: 195–230.

Jasanoff, S. (1990). *The Fifth Branch: Science Advisors as Policymakers*. Cambridge, MA: Harvard University Press.

Jasanoff, S. (Ed.) (1994). *Learning from Disaster: Risk Management after Bhopal*. Philadelphia: University of Pennsylvania Press.

Jasanoff, S. (2003). "Technologies of Humility: Citizen Participation in Governing Science." *Minerva* 41(3): 223–244.

Jasanoff, S. (2005). *Designs on Nature: Science and Democracy in Europe and the United States*. Princeton, NJ: Princeton University Press.

Jasanoff, S., and Kim, S. H. (2015). *Dreamscapes of Modernity: Sociotechnical Imaginaries and the Fabrication of Power*. Chicago: University of Chicago Press.

Jasanoff, S., and Wynne, B. (1998). "Science and Decisionmaking." In Rayner, S., and Malone, E. (Eds.), *Human Choice and Climate Change, Volume 1: The Societal Framework*, 1–88. Pacific Northwest Labs, Columbus: Battelle Press.

Jensen, L., and Yutko, B. (2014). "Why Budget Airlines Could Soon Charge You to Use the Bathroom." *FiveThirtyEight*, June 30. http://fivethirtyeight.com/features/if-everyone-went-to-the-bathroom-before-boarding-the-plane-ticket-prices-might-be-lower/ (accessed September 9, 2016).

Johnson, A. (2001). "Unpacking Reliability: The Success of Robert Bosch, GMBH in Constructing Antilock Braking Systems as Reliable Products." *History and Technology* (17): 249–270.

Jones, R. (1999). *Mechanics of Composite Materials*. 2nd ed. New York: Brunner-Routledge.

Jones-Imhotep, E. (2000). "Disciplining Technology: Electronic Reliability, Cold-War Military Culture and the Topside Ionogram." *History and Technology* 17: 125–175.

Jones-Imhotep, E. (2002). "Reliable Humans, Trustworthy Machines: The Material and Social Construction of Electronic Reliability." Society for the History of Technology Conference Paper.

Jones-Imhotep, E. (2017). *The Unreliable Nation: Hostile Nature and Technological Failure in the Cold War*. Cambridge, MA: MIT Press.

Johnson, S. (2002). *The Secret of Apollo: Systems Management in American and European Space Programs*. Baltimore: Johns Hopkins University Press.

Kahn, L.H. (2011). "Is the United States Prepared for a Nuclear Reactor Accident?" *Bulletin of the Atomic Scientists*, April 7.

Kahneman, D. (2000). "Evaluation by Moments, Past and Future." In Kahneman, D., and Tversky, A. (Eds.), *Choices, Values and Frames*, 693–708. Cambridge: Cambridge University Press.

Kahneman, D. (2011). *Thinking, Fast and Slow*. London: Penguin.

Kaldor, M. (1981). *The Baroque Arsenal*. New York: Hill and Wang.

Kan, N. (2017). *My Nuclear Nightmare: Leading Japan through the Fukushima Disaster to a Nuclear-Free Future*. Translated by Jeffrey S. Irish. Ithaca, NY: Cornell University Press.

Kaplan, F. (2016). *Dark Territory: The Secret History of Cyber War*. London: Simon & Schuster.

Kazel, R. (2015). "Ex-Chief of Nuclear Forces General Lee Butler Still Dismayed by Deterrence Theory and Missiles on Hair Trigger Alert." *Nuclear Age Peace Foundation*, May. https://www.wagingpeace.org/general-lee-butler/ (accessed September 9, 2021).

Kececioglu, D. and Tian, X. (1984). 'Reliability Education: A Historical Perspective." *IEEE Transactions on Reliability* R-33(1): 390–398.

Kemp, R., Schot, J., and Hoogma, R. (1998). "Regime Shifts to Sustainability through Processes of Niche Formation: The Approach of Strategic Niche Management." *Technological Analysis & Strategic Management* 10(2): 175–198.

Khurana, K. C. (2009). *Aviation Management: Global Perspectives*. New Delhi: Global India Publications.

King, M. (2016). *The End of Alchemy: Money, Banking, and the Future of the Global Economy*. London: Norton.

Kirk, S. (1995). *First in Flight: The Wright Brothers in North Carolina*. Winston-Salem, NC: John F. Blair.

Kitroefff, N. and Gelles, D. (2019). "At Boeing, C.E.O.'s Stumbles Deepen a Crisis." *The New-York Times*. December 22, 2019. https://www.nytimes.com/2019/12/22/business/boeing-dennis-muilenburg-737-max.html (accessed March 16, 2023).

Klesius, M. (2009). "Are Aft-Facing Airplane Seats Safer?" *Airspacemag*. https://www.airspacemag.com/need-to-know/are-aft-facing-airplane-seats-safer-146695292/?all (accessed May 2, 2018).

Knight J., and Leveson, N. (1986). "An Experimental Evaluation of the Assumption of Independence in Multiversion Programming." *IEEE Transactions on Software Engineering* 12(1): 96–109.

Komite Nasional Keselamatan Transportasi Republic of Indonesia (KNKT). (2015). "PT. Indonesia Air Asia Airbus A320-216; PK-AXC Karimata Strait Coordinate 3°37′ 19″S-109°42′41″E Republic of Indonesia 28 December 2014." Final Report. http://kemhubri.dephub.go.id/knkt/ntsc_home/ntsc.htm (accessed February 12, 2015).

Komite Nasional Keselamatan Transportasi Republic of Indonesia (KNKT). (November 28, 2018). "Preliminary Aircraft Accident Investigation Report: PT. Lion Mentari Airlines Boeing 737-8 (MAX)." KNKT.18.10.35.04. http://knkt.dephub.go.id/knkt/ntsc_aviation/baru/pre/2018/2018%20-%20035%20-%20PK-LQP%20Preliminary%20 Report.pdf (accessed April 5, 2019).

Komons, N. (1978). *Bonfires to Beacons: Federal Civil Aviation Policy under the Air Commerce Act 1926–38*. Washington, DC: US Government Printing Office.

Kreisher, O. (February 2010). "The Aircraft Losses Mount." *Air Force Magazine*. http://www.airforcemag.com/MagazineArchive/Pages/2010/February%202010/0210aircraft .aspx (accessed March 12, 2019).

Kubiak, W. D. (2011). "Fukushima's Cesium Spew—Deadly Catch-22s in Japan Disaster Relief." *Truthout*, June 27. http://www.truth-out.org/fukushimas-cesium-spew-eludes -prussian-blues-deadly-catch-22s-japan-disaster-relief/1308930096 (accessed June 19, 2013).

Kuhn, T. (1996 [1962]). *The Structure of Scientific Revolutions,* 3rd ed. Chicago: University of Chicago Press.

Kusch, M. (2012). "Sociology of Science: Bloor, Collins, Latour." In Brown, J. (Ed.), *Philosophy of Science: The Key Thinkers.* London: Continuum Books: 165–182.

Lagoni, N. I. (2007). *The Liability of Classification Societies.* Berlin: Springer.

Lampland, M. (2010). "False Numbers as Formalizing Practices." *Social Studies of Science* 40(3): 377–404.

Langewiesche, W. (1998a). *Inside the Sky: A Meditation on Flight.* New York: Pantheon.

Langewiesche, W. (1998b). "The Lessons of ValuJet 592." *Atlantic Monthly,* March. http://www.theatlantic.com/magazine/archive/1998/03/the-lessons-of-valujet-592 /306534/ (accessed September 4, 2017).

Langewiesche, W. (2003). "Columbia's Last Flight." *The Atlantic,* November. https:// www.theatlantic.com/magazine/archive/2003/11/columbias-last-flight/304204/ (accessed September 4, 2017).

Langewiesche, W. (2009a). "Anatomy of a Miracle." *Vanity Fair,* June.

Langewiesche, W. (2009b). *Fly by Wire: The Geese, the Glide, the Miracle on the Hudson.* New York: Farrar, Straus & Giroux.

Langewiesche, W. (2014). "The Human Factor." *Vanity Fair,* October. http://mg.co .za/article/2010-02-22-the-human-factor (accessed December 1, 2015).

LaPorte, T. (1982). "On the Design and Management of Nearly Error-Free Organizational Control Systems." In Sills, D., Shelanski, V., and Wolf. C. (Eds.), *Accident at Three Mile Island: The Human Dimensions,* 185–200. Boulder, CO: Westview.

LaPorte, T. (1994). "A Strawman Speaks Up: Comments on the Limits of Safety." *Journal of Contingencies and Crisis Management* 2(2): 207–212.

LaPorte, T., and Consolini, P. M. (1991). "Working in Practice but Not in Theory: Theoretical Challenges of 'High Reliability Organizations'." *Journal of Public Administration Research and Theory* 1(1): 19–47.

LaPorte, T., and Rochlin, G. (1994). "A Rejoinder to Perrow." *Journal of Contingencies and Crisis Management* 2(4): 221–227.

Latour, B. (1987). *Science in Action: How to Follow Scientists and Engineers through Society.* Cambridge, MA: Harvard University Press.

Latour, B. (1996). *Aramis, or the Love of Technology.* Cambridge, MA: Harvard University Press.

Latour, B. (1999). *Pandora's Hope: Essays on the Reality of Science Studies.* Cambridge, MA: Harvard University Press.

Latour, B. (2004). "Why Has Critique Run out of Steam? From Matters of Fact to Matters of Concern." *Critical Inquiry* 30(2): 225–248.

Latour, B., and Woolgar, S. (1979). *Laboratory Life: The Social Construction of Scientific Facts.* Beverly Hills, CA: SAGE.

Lauber, J. K. (1989). "Human Performance and Aviation Safety: Some Issues and Some Solutions." *Accident Prevention Bulletin* 46(4): 10–11.

Lazonick, W., and O'Sullivan, M. (2000). "Maximising Shareholder Value: A New Ideology for Corporate Governance." *Economy and Society* 29(1): 13–35.

Lean, G. (2012). "Why Nuclear Is in Meltdown." *The Telegraph,* March 2. http://www.telegraph.co.uk/comment/9118831/Why-nuclear-is-in-meltdown.html (accessed August 9, 2013).

Le Coze, J. C. (2015). "1984–2014. 'Normal Accidents': Was Charles Perrow Right for the Wrong Reasons?" *Journal of Contingencies and Crisis Management* 23(4): 275–286.

Le Coze, J. C. (Ed.) (2020). *New Perspectives on Safety Research.* Boca Raton, FL: CDC Press.

Lederer, J. (1968). "Risk speculations of the Apollo Project." *Paper presented at the Wings Club.* New York, December 18, 1968.

Lee, J., Veloso, F. M., Hounshell, D. A., and Rubin, E. S. (2010). "Forcing Technological Change: A Case of Automobile Emissions Control Technology Development in the US." *Technovation* 30: 249–264.

Lenorovitz, J. (1991). "Airbus Survey Confirms Requirement for Very Large Transport Aircraft." *Aviation Week and Space Technology*, October 28.

Leveson, N. (1988). "Airbus Fly-by-Wire Controversy." *The Risks Digest* 6(32). Friday, February 26, 1988. http://catless.ncl.ac.uk/Risks/6.32.html (accessed July 20, 2015).

Leveson, N., Dulac, N., Marais, K., and Carroll, J. (2009). "Moving beyond Normal Accidents and High-Reliability Organizations: A Systems Approach to Safety in Complex Systems." *Organization Studies* 30(2–3): 227–249.

Lewis, H. (1990). *Technological Risk.* New York: Norton.

Li, V. (2018). "The Next Financial Crisis: Why It Is Looking Like History May Repeat Itself." CNBC, September 14. https://www.cnbc.com/2018/09/14/the-next-financial-crisis-why-it-looks-like-history-may-repeat-itself.html (accessed July 14, 2021).

Littlewood, B. (1996). "The Impact of Diversity upon Common Cause Failure." *Reliability Engineering & System Safety* 51(1): 101–113.

Littlewood, B., and Miller, D. (1989). "Conceptual Modeling of Coincident Failures in Multi-Version Software." *IEEE Transactions on Software Engineering* 15(12): 1596–1614.

Littlewood, B., Popov, P., and Strigini, L. (1999). "A Note on the Reliability Estimation of Functionally Diverse Systems." *Reliability Engineering & System Safety* 66: 93–95.

Littlewood, B., Popov, P., and Strigini, L. (2002). "Assessing the Reliability of Diverse Fault-Tolerant Systems." *Safety Science* 40: 781–796.

Littlewood, B., and Strigini, L. (1993). "Validation of Ultra-High Dependability for Software-based Systems." *Communications of the ACM* 36(11): 69–80.

Littlewood, B., and Wright, D. (1997). "Some Conservative Stopping Rules for the Operational Testing of Safety-Critical Software." *IEEE Transactions on Software Engineering* 23(11): 673–683.

Lloyd, E., and Tye, W. (1982). *Systematic Safety: Safety Assessment of Aircraft Systems.* London: Civil Aviation Authority (CAA).

Lochbaum, D., Lyman, E., and Stranahan, S. (2014). *Fukushima: The Story of a Nuclear Disaster.* New York: New Press.

Loftin, L. (1985). *Quest for Performance: The Evolution of Modern Aircraft.* Washington, DC: NASA Scientific and Technical Information Branch.

Longmate, N. (1985). *Hitler's Rockets: The Story of the V-2s.* London: Hutchinson.

Losey, S. (2018). "Fewer Planes Are Ready to Fly: Air Force Mission-Capable Rates Decline amid Pilot Crisis." *Air Force Times*, March 5. https://www.airforcetimes.com /news/your-air-force/2018/03/05/fewer-planes-are-ready-to-fly-air-force-mission -capable-rates-decline-amid-pilot-crisis/ (accessed March 3, 2019).

Lowenstein, R. (2011). *The End of Wall Street.* London: Penguin.

Lowy, J. (2010). "Cascading Failures Followed Airline Engine Blowout." *San-Diego Union Tribune,* November 10. http://www.sandiegouniontribune.com/sdut-cascading -failures-followed-airline-engine-blowout-2010nov18-story.html (accessed September 24, 2017).

Lowy, J. (2018). "Fact Check: Trump Wrongly Claims Credit for Lack of Commercial Airline Crash Deaths." *Chicago Tribune*, January 2. http://www.chicagotribune.com /news/nationworld/politics/factcheck/ct-trump-airline-safety-fact-check-20180102 -story.html (accessed September 21, 2018).

Luhmann, N. (2005). *Risk: A Sociological Theory.* New Brunswick, NJ: Aldine.

Lundberg, B. K. O. (1965). "The Supersonic Adventure." *Bulletin of the Atomic Scientists*, February: 29–33.

Lynch, M. (2017). "STS, Symmetry and Post-Truth." *Social Studies of Science* 47(4): 593–599.

Lynch, M., and Cole, S. (2002). "Judicial Metascience and the Credibility of Expert Evidence." European Association for the Study of Science & Technology (EASST) Conference, University of York, UK (July 31—August 3, 2002).

Macarthur, J., and Tesch, M. (1999). *Air Disaster, Volume 3.* Fyshwick, Australia: Aerospace Publications.

Macheras, A. (2019). "The Boeing Crisis: One Month Later." *Aviation Analyst,* April 10. https://aviationanalyst.co.uk/2019/04/10/the-boeing-crisis-one-month-later/ (accessed April 8, 2019).

MacKenzie, D. (1989). "From Kwajalein to Armageddon? Testing and the Social Construction of Missile Accuracy." In Gooding, D., Pinch, T. and Schaffer, S. (Eds.), *The Uses of Experiment: Studies in the Natural Sciences*, 409–435. Cambridge: Cambridge University Press.

MacKenzie, D. (1990). *Inventing Accuracy: A Historical Sociology of Nuclear Missile Guidance.* Cambridge, MA: MIT Press.

MacKenzie, D. (1996a). "How Do We Know the Properties of Artifacts? Applying the Sociology of Knowledge to Technology." In Fox, R. (Ed.), *Technological Change:*

Methods and Themes in the History of Technology, 249–251. Amsterdam: Harwood Academic.

MacKenzie, D. (1996b). *Knowing Machines: Essays on Technical Change*. Cambridge, MA: MIT Press.

MacKenzie, D. (2001). *Mechanizing Proof: Computing, Risk, and Trust*. Cambridge, MA: MIT Press.

MacKenzie, D. (2005). *An Engine, not a Camera. How Financial Models Shape Markets*. Cambridge, MA: MIT Press.

MacKenzie, D., and Spinardi G. (1996). "Tacit Knowledge and the Uninvention of Nuclear Weapons." *American Journal of Sociology* 101(1): 44–99.

MacKinnon, B., Sowden, R., and Dudley, S. (Eds.). (2001). *Sharing the Skies: An Aviation Guide to the Management of Wildlife Hazards*. Ottawa: Transport Canada.

Macrae, C. (2007). "Interrogating the Unknown: Risk Analysis and Sensemaking in Airline Safety Oversight." Discussion Paper No: 43. ESRC Centre for Analysis of Risk and Regulation (LSE).

Macrae, C. (2014). *Close Calls: Managing Risk and Resilience in Airline Flight Safety*. London: Palgrave Macmillan.

Maksel, R. (2008). "What Determines an Airplane's Lifespan?" *Air & Space Magazine*, March 1.

March, J., and Olsen, J. (1988). "The Uncertainty of the Past: Organizational Learning under Ambiguity." In March, J., *Decisions and Organizations*. New York: Wiley.

March, J., and Simon, H. (1958). *Organizations*. New York: Wiley.

March, J., Sproull, L., and Tamuz, M. (1991). "Learning from Samples of One or Fewer." *Organization Science* 2(1): 1–13.

Marks, P. (2006). "Warning Signs." *New Scientist* 191(2560). July 12.

Marks, P. (2009). "Why Large Carbon-Fibre Planes Are Still Grounded." *New Scientist*, August 22, 20.

Marra, P. P., Dove, C. J., Dolbeer, R., et al. (2009). "Migratory Canada Geese Cause Crash of US Airways Flight 1549." *Frontiers in Ecology and the Environment* 7(6): 297–301.

Marsh, G. (2009). "Boeing's 787: Trials, Tribulations, and Restoring the Dream." *Reinforced Plastics* November/December: 16–21.

Matsumura, A. (2012). "Fukushima Daiichi Site: Cesium-137 Is 85 Times Greater than at Chernobyl Accident." *Finding the Missing Link,* April 3. https://www.nrc.gov /docs/ML1210/ML12103A214.pdf (accessed March 10, 2023).

McCurdy, H. (1993). *Inside NASA: High Technology and Organizational Change in the U.S. Space Program*. Baltimore: John Hopkins University Press.

McNeill, D., and Adelstein, J. (2011). "Meltdown: What Really Happened at Fukushima?" *Atlantic Wire*, July 2. http://www.theatlanticwire.com/global/2011/07/melt down-what-really-happened-fukushima/39541/ (accessed June 19, 2013).

Mecham, M. (2003). "Betting on Suppliers." *Aviation Week and Space Technology*. October 23, 2003.

Mecham, M. (2005). "More Flight Time: Composites and Electric Systems Should Help Keep Boeing's 787 Out of Maintenance Shops." *Aviation Week and Space Technology*, April 18.

Mecham, M. (2009). "Starts and Fits." *Aviation Week and Space Technology*, June 29, 2009.

Merlin, P. (2009). "Design and Development of the Blackbird: Challenges and Lessons Learned." Paper delivered at 47th AIAA Aerospace Sciences Meeting, Including the New Horizons Forum and Aerospace Exposition January 5–8, 2009, Orlando, FL.

Merton, R. K. (1936). "The Unanticipated Consequences of Purposive Social Action." *American Sociological Review* 1(6): 894–904.

Merton, R. K. (1940). "Bureaucratic Structure and Personality." *Social Forces* 18: 560–568.

Mihm, S. (2019). "The FAA Has Always Played Cozy with the Aviation Industry." *Bloomberg Opinion*, March 21. https://www.bloomberg.com/opinion/articles/2019-03-21/boeing-737-crash-faa-confronts-its-confusing-role-as-regulator (accessed March 27, 2019).

Miller, C. R. (2003a). "The Presumptions of Expertise: The Role of Ethos in Risk Analysis." *Configurations* 11(2), 163–202.

Miller, P. (2003b). "Governing by Numbers: Why Calculative Practices Matter." In *The Blackwell Cultural Economy Reader*, 179–189. London: Blackwell.

Miller, R., and Sawers, D. (1970). *The Technical Development of Modern Aviation*. New York: Praeger.

Miller, P., and Napier, C. (1993). "Genealogies of Calculation." *Accounting, Organizations and Society* 18(7–8): 631–647.

Minoura, K., Imamura, F., Sugawara, D., Kono, Y. and Iwashita, T. (2001). "The 869 Jogan Tsunami Deposit and Recurrence Interval of Large-Scale Tsunami on the Pacific Coast of Northeast Japan." *Journal of Natural Disaster Science* 23: 83–88.

Mitcham, C. (1994). *Thinking through Technology: The Path between Engineering and Philosophy*. Chicago: University of Chicago Press.

Mohney, G. (2014). "Long Search for Missing Plane Could Cost 'Hundreds of Millions of Dollars.'" ABC News Online, March 13. http://abcnews.go.com/International/long-search-missing-plane-cost-hundreds-millions-dollars/story?id=22899690 (accessed December 16, 2015).

Mokyr, J. (1990). *The Lever of Riches: Technological Creativity and Economic Progress*. Oxford: Oxford University Press.

Moon, H. (1989). *Soviet SST: The Technopolitics of the Tupolev-144*. London: Orion Books.

Morris, H. (2017). "58,000 Miles and 46 Flights: A Week in the Extraordinary Life of a Modern Aircraft." *The Telegraph*, October 13.

Mowery, D. C., and Rosenberg, N. (1981). "Technical Change in the Commercial Aircraft Industry, 1925–1975." *Technological Forecasting and Social Change* 20: 347–358.

Munro, E. (2004). "The Impact of Audit on Social Work Practice." *British Journal of Social Work* 34(8): 1075–1095.

Murawski, S. A., Hollander, D. J., Gilbert, S. and Gracia, A. (2020). "Deepwater Oil and Gas Production in the Gulf of Mexico and Related Global Trends." In Murawski, S., Ainsworth, C. H., Gilbert, S., et al. (Eds.), *Scenarios and Responses to Future Deep Oil Spills*, 16–33. Cham, Switzerland: Springer.

Musso, C. (2009). "New Learning from Old Plastics: The Effects of Value-Chain-Complexity on Adoption Time." *Technovation* 29(4): 299–312.

Myre, G. (2017) "Stanislav Petrov, 'The Man Who Saved the World,' Dies at 77." NPR, September 18. https://www.npr.org/sections/thetwo-way/2017/09/18/551792129/stanislav-petrov-the-man-who-saved-the-world-dies-at-77?t=1628523159902 (accessed September 9, 2021).

Nader, R., and Smith, W. (1994). *Collision Course: The Truth about Airline Safety*. New York: TAB Books.

National Academy of Sciences (NAS) (National Research Council). (1980). *Improving Aircraft Safety: FAA Certification of Commercial Passenger Aircraft*. Washington, DC: National Academies Press.

National Academy of Sciences (NAS) (National Research Council). (1998). *Improving the Continued Airworthiness of Civil Aircraft: A Strategy for the FAA's Aircraft Certification Service*. Washington, DC: National Academies Press.

National Commission on the BP Deepwater Horizon Oil Spill and Offshore Drilling (NCBP). (January 2011). *Deep Water: The Gulf Oil Disaster and the Future of Offshore Drilling*. Report to the President.

National Transportation Safety Board (NTSB). (1972). "Aircraft Accident Report: Capitol International Airways, Inc., DC-8-63F, N4909C, Anchorage, Alaska, November 27, 1970." NTSB-ARR-72-12. No. i-0025.

National Transportation Safety Board (NTSB). (1973). "NTSB-AAR-73-02 Report, Aircraft Accident Report: American Airlines, Inc. McDonnell Douglas DC-10-10, N103AA. Near Windsor, Ontario, Canada. June 12, 1972." February 28, 1973. Washington, DC.

National Transportation Safety Board (NTSB). (1989). "Aircraft Accident Report— Aloha Airlines, Flight 243, Boeing 737-200, N73711, near Maui, Hawaii, April 28, 1988." Report no: NTSB/AAR-89/03; Acc. No: PB89–91C404. Washington, DC.

National Transportation Safety Board (NTSB). (1996). "In-Flight Fire and Impact with Terrain ValuJet Airlines Flight 592, DC-9-32, N904VJ. Everglades, near Miami, Florida. May 11, 1996." Report no: NTSB/AAR-97/06; Acc. No: PB97–910406. Washington, DC.

National Transportation Safety Board (NTSB). (2000a). "In-flight Breakup over the Atlantic Ocean, Trans World Airlines Flight 800, Boeing 747-131, N93119, Near East Moriches, New York, July 17, 1996." Aircraft Accident Report NTSB/AAR-00/03. Washington, DC.

National Transportation Safety Board (NTSB). (2000b). Loss of Control and Impact with Pacific Ocean, Alaska Airlines Flight 261, McDonnell Douglas MD-83, N963AS, about 2.7 Miles North of Anacapa Island, California, January 31, 2000. Aircraft Accident Report NTSB/AAR-02/01. Washington, DC.

National Transportation Safety Board (NTSB). (2006a). "Safety Recommendation A-06-27—A-06-28." Washington, DC.

National Transportation Safety Board (NTSB). (2006b). "Safety Report on the Treatment of Safety-Critical Systems in Transport Airplanes." Safety Report NTSB/SR-06/02. PB2006-917003. Notation 7752A. Washington, DC.

National Transportation Safety Board (NTSB). (2009). Loss of Thrust in Both Engines after Encountering a Flock of Birds and Subsequent Ditching on the Hudson River, 213. Accident Report. NTSB/AAR-10/03. Washington, DC.

National Transportation Safety Board (NTSB). (2013). "Auxiliary Power Unit Battery Fire Japan Airlines Boeing 787—8, JA829J." No. NTSB/AIR-14/01. Boston.

Neufeld, M. (1990). "The Guided Missile and the Third Reich: Peenemünde and the Forging of a Technological Revolution." In Renneberg, M. and Walker, M. (Eds.), *Science, Technology, and National Socialism*, 51–71. Cambridge: Cambridge University Press.

Newhouse, J. (1982). *The Sporty Game: The High-Risk Competitive Business of Making and Selling Commercial Airliners*. New York: Knopf.

Neyland, D., and Woolgar, S. (2002). "Accountability in Action? The Case of a Database Purchasing Decision." *British Journal of Sociology* 53: 259–274.

Nicas, J., and Creswell, J. (2019). "Boeing's 737 Max: 1960s Design, 1990s Computing Power and Paper Manuals." *New York Times*, April 8. https://www.nytimes.com/2019/04/08/business/boeing-737-max-.html (accessed April 8, 2019).

Nicas, J., Kitroeff, N., Gelles, D. and Glanz, J. (2019). "Boeing Built Deadly Assumptions Into 737 Max, Blind to a Late Design Change" *New York Times*, June 1.

Niles, M. (2002). "On the Hijacking of Agencies (and Airplanes): The Federal Aviation Administration. 'Agency Capture' and Airline Security." *Journal of Gender, Social Policy & the Law* 10(2): 381–442.

Nöggerath, J., Geller, R., and Gusiakov, V. (2011). "Fukushima: The Myth of Safety, the Reality of Geoscience." *Bulletin of the Atomic Scientists* 67(5): 37–46.

Norris, G. (2009). "Flying Start." *Aviation Week and Space Technology*, December 21, 2009.

Norris, G., and Hills, B. (1994). "Transcript Reveals Cockpit Anarchy." *Flight International* October, 5–11.

Norris, G., and Wagner, M. (1999). *Airbus*. St. Paul, MN: MBI Publishing.

Nuclear Energy Institute (NEI). (2016). "World Statistics." http://www.nei.org/Knowledge-Center/Nuclear-Statistics/World-Statistics (accessed August 1, 2016).

Oberstar, J. L. and Mica, J. L. (2008). "Critical Lapses in FAA Safety Oversight of Airlines: Abuses of Regulatory Partnership Programs." Summary of Subject Matter, US House of Representatives Committee on Transportation and Infrastructure. Washington, DC: US Government Printing Office.

Office of Technology Assessment (OTA). (July 1988). "Safe Skies for Tomorrow: Aviation Safety in a Competitive Environment." OTA-SET-381. Washington, DC: US Government Printing Office.

O'Hehir, A. (2016). "The Night We Almost Lost Arkansas—A 1980 Nuclear Armageddon That Almost Was." *Salon*, September 14. https://www.salon.com/2016/09/14 /the-night-we-almost-lost-arkansas-a-1980-nuclear-armageddon-that-almost-was/ (accessed September 9, 2021).

Okrent, D. (1978). *On the History of the Evolution of Light Water Reactor Safety in the United States*. International Panel on Fissile Materials. http://fissilematerials.org /library/1978/06/on_the_history_of_the_evolutio.html (accessed 07/05/2012).

Onishi, N. (2011). "'Safety Myth' Left Japan Ripe for Nuclear Crisis." *New York Times*, June 24.

Onishi, N., and Fackler, M. (2011). "Japan Held Nuclear Data, Leaving Evacuees in Peril." *New York Times*, August 8.

Oreskes, N., and Conway, E. (2010). *Merchants of Doubt: How a Handful of Scientists Obscured the Truth on Issues from Tobacco Smoke to Global Warming*. London: Bloomsbury.

Orlady, H. (2017). *Human Factors in Multi-Crew Flight Operations*. London: Routledge.

Orlebar, C. (1997). *The Concorde Story*. Oxford: Osprey.

O'Rourke, R. (2009). "Air Force F-22 Fighter Program: Background and Issues for Congress." *CRS Report for Congress* RL31673. July 16. Washington, DC: Congressional Research Service.

Osnos E. (2011). "The Fallout. Seven Months Later: Japan's Nuclear Predicament." *The New Yorker*, October 17, 26–61.

Ostrower, J. (2011). "Boeing Aims to Minimise 737 Max Changes." *Air Transport Intelligence News*, August 31. https://www.flightglobal.com/news/articles/boeing -aims-to-minimise-737-max-changes-361440/ (accessed April 5, 2019).

Owen, K. (2001). *Concorde: Story of a Supersonic Pioneer*. London: Science Museum.

Partnoy, F. (2020). "The Looming Bank Collapse." *The Atlantic*, July/August. https:// www.theatlantic.com/magazine/archive/2020/07/coronavirus-banks-collapse /612247/ (accessed September 14, 2021).

Peltzman, S. (1976). "Toward a More General Theory of Regulation." *Journal of Law and Economics* 19(2): 211–240.

Perin, C. (2005). *Shouldering Risks: The Culture of Control in the Nuclear Power Industry*. Princeton, NJ: Princeton University Press.

Perrow, C. (1983). "The Organizational Context of Human Factors Engineering." *Administrative Science Quarterly* 28(4): 521–541.

Perrow, C. (1984). *Normal Accidents: Living with High-Risk Technologies*. New York: Basic Books.

Perrow, C. (1994). "The Limits of Safety: The Enhancement of a Theory of Accidents." *Journal of Contingencies and Crisis Management* 4(2): 212–220.

Perrow, C. (1999). *Normal Accidents: Living with High-Risk Technologies* 2nd ed. Princeton, NJ: Princeton University Press.

Perrow, C. (2007). *The Next Catastrophe: Reducing Our Vulnerabilities to Natural, Industrial, and Terrorist Disasters*. Princeton, NJ: Princeton University Press.

Perrow, C. (2011). "Fukushima and the Inevitability of Accidents." *Bulletin of the Atomic Scientists* 67(6): 44–52.

Perrow, C. (2015). "Cracks in the 'Regulatory State.'" *Social Currents* 2(3): 203–212.

Petroski, H. (1992a). *To Engineer Is Human: The Role of Failure in Successful Design*. New York: Vintage Books.

Petroski, H. (1992b). *The Evolution of Useful Things: How Everyday Artifacts from Forks and Pins to Paperclips and Zippers Came to Be as They Are*. New York: Alfred A. Knopf.

Petroski, H. (1994). *Design Paradigms: Case Histories of Error and Judgment in Engineering*. Cambridge: Cambridge University Press.

Petroski, H. (2008). *Success through Failure: The Paradox of Design*. Princeton, NJ: Princeton University Press.

Philips, A. (1998). "20 Mishaps That Might Have Started Accidental Nuclear War." Nuclearfiles.org. https://web.archive.org/web/20200703203219/http://www.nuclearfiles .org/menu/key-issues/nuclear-weapons/issues/accidents/20-mishaps-maybe-caused -nuclear-war.htm (accessed September 9, 2021).

Pilkington, E. (2013). "US Nearly Detonated Atomic Bomb over North Carolina— Secret Document." *The Guardian*, September 20. https://www.theguardian.com /world/2013/sep/20/usaf-atomic-bomb-north-carolina-1961 (accessed September 9, 2021).

Pinch, T. (1991). "How Do We Treat Technical Uncertainty in Systems Failure? The Case of the Space Shuttle Challenger." In LaPorte, T. (Ed.), *Social Responses to Large Technical Systems: Control or Anticipation*, 143–158. Dordrecht, Netherlands: Kluwer Academic Publishers.

Pinch, T. (1993). "'Testing—One, Two, Three . . . Testing!': Toward a Sociology of Testing." *Science, Technology, & Human Values* 18(1): 25–41.

Pinch, T., and Bijker, W. (1984). "The Social Construction of Facts and Artifacts: or How the Sociology of Science and the Sociology of Technology Might Benefit Each Other." *Social Studies of Science* 14: 339–441.

Polanyi, M. (1958). *Personal Knowledge*. London: Routledge & Kegan Paul.

Pollock, N., and Williams, R. (2010). "The Business of Expectations: How Promissory Organizations Shape Technology and Innovation." *Social Studies of Science* 40(4): 525–548.

Pope, S. (2008). "Fly by Wire Technology." *AINonline*, February 8. https://www .ainonline.com/aviation-news/aviation-international-news/2008-02-08/fly-wire -technology (accessed September 6, 2015).

Popov, P., Strigini, L., May, J. and Kuball, S. (2003). "Estimating Bounds on the Reliability of Diverse Systems." *IEEE Transactions on Software Engineering* 29(4): 345–359.

Popper, K. (1959). *The Logic of Scientific Discovery*. New York: Basic Books.

Porter, T. M. (1994). "Making Things Quantitative." In Power, M. (Ed.), *Accounting and Science: Natural Inquiry and Commercial Reason*, 36–57. Cambridge: Cambridge University Press.

Porter, T. M. (1995). *Trust in Numbers: The Pursuit of Objectivity in Scientific and Public Life*. Princeton, NJ: Princeton University Press.

Posner, R. A. (1971). "Taxation by Regulation." *Bell Journal of Economics and Management Science* 2(22): 22–50.

Posner, R. A. (1974). "Theories of Economic Regulation." *Bell Journal of Economics and Management Science* 5(2): 335–358.

Posner, R. A. (1975). "The Social Costs of Monopoly and Regulation." *Journal of Political Economy* 83(4): 807–827.

Power, M. (1997). *The Audit Society: Rituals of Verification*. Oxford: Oxford University Press.

Power, M. (2003). "The Operational Risk Game." *Risk & Regulation* (5). London: LSE Center for Analysis of Risk and Regulation.

Power, M. (2007). *Organized Uncertainty: Designing a World of Risk Management*. Oxford: Oxford University Press.

Proctor, R. (1991). *Value-Free Science?: Purity and Power in Modern Knowledge*. Cambridge, MA: Harvard University Press.

Quine, W. V. O. (1975). "On Empirically Equivalent Systems of the World." *Erkenntnis* 9(3): 313–328.

Quintana, M. (2012). "Fukushima Crisis Concealed: Japanese Government Kept Worst-Case Scenario under Wraps." *Asia Pacific Journal*, January 31, 2012. http://japanfocus.org///events/view/129 (accessed September 8, 2013).

Rae. J. B. (1968). *Climb to Greatness: The American Aircraft Industry, 1920–1960*. Cambridge, MA: MIT Press.

Raju, S. (2016). "Estimating the Frequency of Nuclear Accidents." *Science & Global Security* 24(1): 37–62.

Raman, R., Graser, J., and Younossi, O. (2003). *The Effects of Advanced Materials on Airframe Operating and Support Costs*. RAND Documented Briefing. Santa Monica, CA: RAND.

Ramana, M. V. (2011). "Beyond Our Imagination: Fukushima and the Problem of Assessing Risk." *Bulletin of the Atomic Scientists*, April 19. https://thebulletin.org/2011/04/beyond-our-imagination-fukushima-and-the-problem-of-assessing-risk/ (accessed March 7, 2023).

Rasmussen, J. (1983). Human Error. *Position Paper for NATO Conference on Human Error*. August 1983, Bellagio, Italy.

Rasmussen, J. (1990). "Human Error and the Problem of Causality in the Analysis of Accidents." *Philosophical Transactions of the Royal Society of London* B327: 449–462.

Rasmussen, J. (1997). "Risk Management in a Dynamic Society: A Modeling Problem." *Safety Science* 27(2): 183–213.

Reason, J. (1990). *Human Error*. Cambridge: Cambridge University Press.

Reason, J. (1997). *Managing the Risks of Organisational Accidents*. London: Ashgate.

Reason, J. (2016). *Organizational accidents revisited*. Aldershot, UK: Ashgate.

Regulinski, T. (1984). "One Score and Fifteen Years Ago." *IEEE Transactions on Reliability* R-33(1): 65–67.

Rhodes, J. (July 1990). "The Black Jet." *Air Force Magazine*, Air Force Association, 73(7). https://www.airandspaceforces.com/article/0790blackjet/ (accessed March 10, 2023).

Rijpma, J. (1997). "Complexity, Tight-Coupling and Reliability: Connecting Normal Accidents Theory and High Reliability Theory." *Journal of Contingencies and Crisis Management* 5(1): 15–23.

Rip, A. (1985). "Experts in Public Arenas." In Otway, H. and Peltu, M. (Eds.), *Regulating Industrial Risks: Science Hazards and Public Protection*, 94–110. London: Butterworths.

Rip, A. (1986). "The Mutual Dependence of Risk Research and Political Context." *Science & Technology Studies* 4(3/4): 3–15.

Risk Assessment Review Group (RARG). (September 1978). *Risk Assessment Review Group Report to the U.S. Nuclear Regulatory Commission*. NUREG/CR-0400.

Roberts, K. H. (1989). "New Challenges in Organization Research: High Reliability Organizations." *Industrial Crisis Quarterly* 3(2): 111–125.

Roberts, K. H. (1993). "Introduction." In Roberts, K. H. (Ed.), *New Challenges to Understanding Organizations*, 1–10. New York: Macmillan.

Robison, P., and Newkirk, M. (2019). "Relationship between Boeing, FAA Safety Regulators under Scrutiny." *Insurance Journal,* March 25. https://www.insurancejournal.com/news/national/2019/03/25/521514.htm (accessed March 27, 2019).

Rochlin, G. I., LaPorte, T. R., and Roberts, K. H. (1987). "The Self-Designing High-Reliability Organization: Aircraft Carrier Flight Operations at Sea." *Naval War College Review* 40(4): 76–90.

Rogers, J. (1996). *Advanced Composite Materials: The Air Force's Role in Technology Development*. RAND Corporation, Document Number: N-3503-AF.

Rothstein, H., and Downer, J. (2012). "Renewing Defra: Exploring the Emergence of Risk-Based Policymaking in UK Central Government." *Public Administration.* 90(3): 781–799.

Rothstein, H., Huber, M., and Gaskell, G. (2006). "A Theory of Risk Colonization: The Spiraling Regulatory Logics of Societal and Institutional Risk." *Economy and Society* 35(1): 91–112.

Rozell, N. (1996). "The Boeing 777 Does More with Less." *Alaska Science Forum,* May 23. https://www.gi.alaska.edu/alaska-science-forum/boeing-777-does-more-less (accessed March 10, 2023).

Rushby, J. (December 1993). *Formal Methods and the Certification of Critical Systems*. Technical Report CSL-93-7.

Saba, J. (1983). "Aircraft Crashworthiness in the United States: Some Legal and Technical Parameters." *Journal of Air Law and Commerce* 48(2): 287–346.

SAE International. (1996). "ARP4761: Guidelines and Methods for Conducting the Safety Assessment Process on Civil Airborne Systems and Equipment." January 12. Warrendale, PA.

Sagan, S. (1993). *The Limits of Safety*. Princeton, NJ: Princeton University Press.

Sagan, S. (2004). "The Problem of Redundancy Problem: Why More Nuclear Security Forces May Produce Less Nuclear Security." *Risk Analysis* 24(4): 935–946.

Salter, L. (1988). *Mandated Science: Science and Scientists in the Making of Standards*. Dordrecht, Netherlands: Springer.

Saxon, W. (1994) "Paul K. Feyerabend, 70, Anti-Science Philosopher." *New York Times*. March 8.

Schlosser, E. (2013). *Command and Control: Nuclear Weapons, the Damascus Accident, and the Illusion of Safety*. New York: Allen Lane.

Schmidt, J. (2009). "The Definition of Structural Engineering." *Structure Magazine*, January: 9.

Schulman, P. (1993). "The Negotiated Order of Organizational Reliability." *Administration & Society* 25(3): 353–372.

Scott, J. (1998). *Seeing like a State: How Certain Schemes to Improve the Human Condition Have Failed*. New Haven, CT: Yale University Press.

Schiavo, M. (1997). *Flying Blind, Flying Safe*. New York: Avon Books.

Shalal-Esa, A. (2014). "Exclusive: Pentagon Report Faults F-35 on Software, Reliability." *Reuters*, January 24. https://www.reuters.com/article/us-lockheed-fighter-exclusive /exclusive-pentagon-report-faults-f-35-on-software-reliability-idUSBREA0N0ID20140124 (accessed March 14, 2019).

Shanahan, D. (2004). "Basic Principles of Crashworthiness." NATO RTO-EN-HFM-113. https://www.semanticscholar.org/paper/Basic-Principles-of-Crashworthiness -Shanahan/225b4ff4b954467a26708eb596c3cdbcd304b5db (accessed March 10, 2023).

Shapin, S. (1995). "Cordelia's Love: Credibility and the Social Studies of Science." *Perspectives on Science* 3: 266–268.

Shatzberg, E. (1999). *Wings of Wood, Wings of Metal: Culture and Technical Choice in American Airplane Materials, 1914–1945*. Princeton, NJ: Princeton University Press.

Shepardson, D. (2019). "FAA must ramp up staffing to oversee airplane certification after 737 MAX-panel." *Reuters*. October 11, 2019. https://www.nasdaq.com/articles /faa-must-ramp-up-staffing-to-oversee-airplane-certification-after-737-max-panel -2019-10-11 (accessed March 11, 2023).

Shewhart, W. (1931). *Economic Control of Manufactured Product*. New York: D. van Nostrand.

Shinners, S. (1967). *Techniques of System Engineering*. New York: McGraw-Hill.

Shivastava, P. (1987). *Bhopal: Anatomy of a Crisis*. Cambridge, MA: Ballinger.

Shrader-Frechette, K. (1980). "Technology Assessment as Applied Philosophy of Science." *Science, Technology, & Human Values* 6(33): 33–50.

Sieg, L., and Kubota, Y. (2012). "Nuclear Crisis Turns Japan Ex-PM Kan into Energy Apostle." *Reuters*, February 17, 2012. http://www.reuters.com/article/2012/02/17/us-japan-kan-idUSTRE81G08P20120217 (accessed September 7, 2015).

Silbey, S. (2009). "Taming Prometheus: Talk about Safety and Culture." *Annual Review of Sociology* 35(1): 341–369.

Simons, G. M. (2012). *Concorde Conspiracy: The Battle for American Skies 1962–77*. London: The History Press.

Sims, B. (1999). "Concrete Practices: Testing in an Earthquake-Engineering Laboratory." *Social Studies of Science* 29(4): 483–518.

Sismondo, S. (2017). "Post-Truth?" *Social Studies of Science* 47(1): 3–6.

Sismondo, S. (2010). *An Introduction to Science and Technology Studies*, 2nd ed. Chichester, UK: Blackwell.

Sjöberg, L. (2004). "Local Acceptance of a High-Level Nuclear Waste Repository." *Risk Analysis* 24(3): 737–749.

Slayton, R., and Spinardi, G. (2016). "Radical Innovation in Scaling Up: Boeing's Dreamliner and the Challenge of Socio-Technical Transitions." *Technovation* 47: 47–58.

Slovic, P. (2012). "The Perception Gap: Radiation and Risk." *Bulletin of the Atomic Scientists* 68(3): 67–75.

Smith, J. (1981). "FAA Is Cool to Cabin Safety Improvements." *Science* 211(4482): 557–560.

Smith, G., and Mindell, D. (2000). "The Emergence of the Turbofan Engine." In Galison, P. and Roland, A. (Eds.), *Atmospheric Flight in the Twentieth Century*, 107–155. Boston: Kluwer.

Smith, J. (2009). "High-Priced F-22 Fighter Has Major Shortcomings." *Washington Post*, July 10.

Smith, O. (2013). "Rear-Facing Aircraft Seats 'Safer.'" *The Telegraph*, July 10, 2013.

Smith, O. (2017). "13 Unbelievable Statistics about Air Travel." *The Telegraph*, September 20. https://www.telegraph.co.uk/travel/lists/surprising-things-about-air-travel/ (accessed September 27, 2018).

Snook, S. (2000). *Friendly Fire*. Princeton, NJ: Princeton University Press.

Snyder, R. (1982). "Impact Protection in Air Transport Passenger Seat Design." *SAE Transactions* 91(4): 4312–4337.

Soble, J. (2014). "Beware the Safety Myth Returning to Japan's Nuclear Debate." *Financial Times*, July 13.

Socolow, R. (2011). "Reflections on Fukushima: A Time to Mourn, to Learn, and to Teach." *Bulletin of the Atomic Scientists*, March 21. http://www.thebulletin.org/web-edition/op-eds/reflections-fukushima-time-to-mourn-to-learn-and-to-teach (accessed September 17, 2017).

Spiegelhalter, D. (2017). "Risk and Uncertainty Communication." *Annual Review of Statistics and Its Application* 4: 31–60.

Spinardi, G. (2002). "Industrial Exploitation of Carbon Fibre in the UK, USA and Japan." *Technology Analysis & Strategic Management* 14(4): 381–398

Spinardi, G. (2019). "Performance-Based Design, Expertise Asymmetry, and Professionalism: Fire Safety Regulation in the Neoliberal Era." *Regulation & Governance* 13(4): 520–539.

Srinivasan, T., and Gopi Rethinaraj, T. (2013). "Fukushima and Thereafter: Reassessment of Risks of Nuclear Power." *Energy Policy* 52: 726–736.

Stanford, K. (2009). *Exceeding Our Grasp: Science, History, and the Problem of Unconceived Alternatives*. Oxford: Oxford University Press.

Starr, C. (1969). "Social Benefits Versus Technological Risks." *Science* 165 (3899): 1232–1238

Stevens, M., and Mele, C. (2018). "Causes of False Missile Alerts: The Sun, the Moon and a 46-Cent Chip." https://www.nytimes.com/2018/01/13/us/false-alarm-missile-alerts.html (accessed September 9, 2021).

Stigler, G. J. (1971). "The Theory of Economic Regulation." *Bell Journal of Economics and Management Science* 2(1): 3–21.

Stilgoe, J. (2018). Machine Learning, Social Learning and the Governance of Self-Driving Cars. *Social Studies of Science* 48(1): 25–56.

Stimpson, E., and McCabe, W. (November 2008). "Managing Risks in Civil Aviation." *Aero Safety World*. Flight Safety Foundation.: 10-14. https://flightsafety.org/wp-content/uploads/2016/12/asw_nov08_p10-14.pdf (accessed March 10, 2023).

Stoller, G. (2001). "Engineer Has Alternate Theory on Plane Disaster." *USA Today*, April. http://www.iasa.com.au/folders/Safety_Issues/RiskManagement/alohaagain.html (accessed February 16, 2012).

Strathern, M. (Ed.). (2000). *Audit Cultures: Anthropological Studies in Accountability, Ethics and the Academy*. New York: Routledge.

Sutharshan, B., Mutyala, M., Vijuk, R. P., and Mishra, A. (2011). "The AP1000TM Reactor: Passive Safety and Modular Design." *Energy Procedia* 7: 293–302.

Suvrat, R. (2016). "Estimating the Frequency of Nuclear Accidents." *Science & Global Security*: 37–62.

Swenson, L., Grimwood, J., and Alexander, C. (1998). *This New Ocean: A History of Project Mercury*. Washington, DC: NASA History Office.

Taebi, B. (2017). "Bridging the Gap between Social Acceptance and Ethical Acceptability." *Risk Analysis* 37(10): 1817–1827.

Taebi, B., Roeser, S., and van de Poel, S. (2012). "The Ethics of Nuclear Power: Social Experiments, Intergenerational Justice, and Emotions." *Energy Policy* 202–206.

Tamuz, M. (1987). "The Impact of Computer Surveillance on Air Safety Reporting." *Columbia Journal of World Business* 22: 69–77.

Tamuz, M. (2001). "Learning Disabilities for Regulators." *Administration & Society* 33(3): 276.

Tenney, D., Davis, J., Pipes, R. B., and Johnston, N. (2009). "NASA Composite Materials Development: Lessons Learned and Future Challenges." NATO Research and Technology Agency (RTA) AVT 164—Support of Composite Systems Fall 2009—Bonn.

Thorpe, J. (2003). "Fatalities and Destroyed Civil Aircraft Due to Bird Strikes, 1912–2002" *Paper presented to International Bird Strike Committee.* Warsaw, 5–9 May, online: https://web.archive.org/web/20090227072007/http://www.int-birdstrike.org /Warsaw_Papers/IBSC26 WPSA1.pdf (accessed February 24, 2023).

Tkacik, M. (2019). "Crash Course: How Boeing's Managerial Revolution Created the 737 MAX Disaster." *New Republic*, September 18. https://newrepublic.com/article/154944 /boeing-737-max-investigation-indonesia-lion-air-ethiopian-airlines-managerial -revolution (accessed May 8, 2020).

Tootell, B. (1985). *"All Four Engines Have Failed": The True and Triumphant Story of BA 009 and the "Jakarta Incident."* London: Andre Deutsch.

Topham, G. (2020). "'Designed by Clowns': Boeing Messages Raise Serious Questions about 737 Max." *The Guardian*, January 10. https://www.theguardian.com/business /2020/jan/09/boeing-737-max-internal-messages (accessed August 8, 2020).

Transocean. (2010). "Fleet Specifications: Deepwater Horizon." Archived from original, June 19. https://web.archive.org/web/20100619121120/http://www.deepwater .com/fw/main/Deepwater-Horizon-56C17.html (accessed 04/03/2019).

Transportation Safety Board of Canada (TSB). (2007). *Aviation Investigation Report Loss of Rudder in Flight: Air Transat Airbus A310-308 C-GPAT Miami, Florida, 90 nm S 06 March 2005.* Report Number A05F0047. Minister of Public Works and Government Services Canada 2007 Cat. No. TU3-5/05-2E.

Turner, B. A. (1976). "The Organizational and Interorganizational Development of Disasters." *Administrative Science Quarterly* 21(3): 378–397.

Turner, B. A. (1978). *Man-Made Disasters.* London: Wykeham.

Turner, B. A., and Pidgeon, N.F. (1997). *Man-Made Disasters.* 2nd ed. Oxford, UK: Butterworth-Heinemann.

Twombly, I. (2017). "Fly-by-Wire: The Computer Takes Control." *Flight Training Magazine.* Aircraft Owners and Pilots Association. July 1. https://www.aopa.org/news -and-media/all-news/2017/july/flight-training-magazine/fly-by-wire (accessed April 4, 2019).

Uhlmann, D. (2020). "BP Paid a Steep Price for the Gulf Oil Spill but for the US a Decade Later, It's Business as Usual." *The Conversation.* https://theconversation.com /bp-paid-a-steep-price-for-the-gulf-oil-spill-but-for-the-us-a-decade-later-its-business -as-usual-136905 (accessed July 27, 2020).

Union of Concerned Scientists (UCS). (2015). "Close Calls with Nuclear Weapons." https://www.ucsusa.org/sites/default/files/attach/2015/04/Close%20Calls%20 with%20Nuclear%20Weapons.pdf (accessed September 9, 2021).

Unruh, G. C. (2000). "Understanding Carbon Lock-in." *Energy Policy* 28: 817–830.

US Airways. (February 18, 1999). "Comments on NPRM #FAA-1998-4815." Docket (53265).

US Congress. (1996). "Aviation Safety: Issues Raised by the Crash of Valujet Flight 592." Hearing before the Subcommittee on Aviation, Committee on Transportation and Infrastructure, House of Representatives, One Hundred and Fourth Congress. Second session. June 25, 1996. US Government Printing Office: Washington, DC.

US Department of Defense (DoD). (May 17, 1999). *Human Engineering Program Process and Procedures, MIL-HDBK-46855A*. Washington, DC: US Department of Defense.

US Department of Defense (DoD). (February 10, 2000). *Standard Practice for System Safety*. MIL-STD-882D. Washington, DC: US Department of Defense.

US Department of Justice (DoJ). (January 7, 2021). "Boeing Charged with 737 Max Fraud Conspiracy and Agrees to Pay over $2.5 Billion" (Press release). Washington, DC: US Department of Justice. https://www.justice.gov/opa/pr/boeing-charged-737 -max-fraud-conspiracy-and-agrees-pay-over-25-billion (accessed March 16, 2023).

Useem, J. (2019). "The Long-Forgotten Flight That Sent Boeing off Course." *The Atlantic*, November 20. https://www.theatlantic.com/ideas/archive/2019/11/how-boeing -lost-its-bearings/602188/ (accessed March 27, 2021).

US Environmental Protection Agency (EPA). (2009). *Cancer Risk Coefficients for Environmental Exposure to Radionucleotides*. Federal Guidance Report No.13. EPA 402-R99-001. Washington, DC: US Environmental Protection Agency.

US Federal Aviation Administration (FAA). (1970). "Turbine Engine Foreign Object Ingestion and Rotor Blade Containment Type Certification Procedures." Advisory Circular AC 33-1 B, April 22. Washington, DC: Department of Transportation.

US Federal Aviation Administration (FAA). (1982). "System Design Analysis." Advisory Circular AC 25.1309-1. Washington, DC: Department of Transportation.

US Federal Aviation Administration (FAA). (1988). "System Design and Analysis." Advisory Circular (AC) 25.1309-1A. Washington, DC: Department of Transportation.

US Federal Aviation Administration (FAA). (1998a). "Aircraft Certification Mission Statement." Aircraft Certification Service (AIR) Headquarters Office Home Page. http://www.faa.gov/avr/air/hq/mission.htm. February 10. Washington, DC: Department of Transportation.

US Federal Aviation Administration (FAA). (1998b). "Airworthiness Standards; Bird Ingestion: Notice of Proposed Rulemaking (NPRM). CFR Parts 23, 25 and 33." Docket No. FM-1998-4815; Notice No. 98-18j RIN 21200AF34. Washington, DC: Department of Transportation.

US Federal Aviation Administration (FAA). (1999a). "The FAA and Industry Guide to Product Certification." Notice N8110.80, January 26. Washington, DC: Department of Transportation.

US Federal Aviation Administration (FAA). (1999b) "Guidance for Reviewing Certification Plans to Address Human Factors for Certification of Transport Airplane Flight Decks." Memorandum ANM-99-2. September 29, 1999. Washington, DC: Department of Transportation.

US Federal Aviation Administration (FAA). (2000). "Airworthiness Standards; Bird Ingestion." 14 CFR Parts 23, 25 and 33; Docket No. FAA-1998-4815; Amendment No. 23–54, 25–100, and 33–20. RIN 2120-AF84. Washington, DC: Department of Transportation.

US Federal Aviation Administration (FAA). (2001). "Bird Ingestion Certification Standards." Advisory Circular AC 33.76, January 19. Washington, DC: Department of Transportation.

US Federal Aviation Administration (FAA). (2002a). "Commercial Airplane Certification Process Study." March 2002. Washington, DC: Department of Transportation.

US Federal Aviation Administration (FAA). (2002b). "System Design and Analysis." Advisory Circular (AC) 25.1309-1B. Arsenal Draft. Washington, DC: Department of Transportation.

US Federal Aviation Administration (FAA). (2002c). "System Design and Analysis Harmonization and Technology Update." Aviation Rulemaking Advisory Committee. Draft R6X Phase 1—June 2002. Washington, DC: Department of Transportation.

US Federal Aviation Administration (FAA). (2003). "Identification of Flight Critical System Components." Memorandum ANM-03-117-10. July 24. Washington, DC: Department of Transportation.

US Federal Aviation Administration (FAA). (2004). "Response to Comments on NPA-E-20." http://www.jaa.nl/section1/crd/crd%20for%20npa%20e-20.doc (accessed March 7, 2015).

US Federal Aviation Administration (FAA). (2005a). *Designee Management Handbook.* FAA Order 8100.8B. July 14. Washington, DC: Department of Transportation.

US Federal Aviation Administration (FAA). (2005b). "Type Certification." Order 8110.4C. October 26. Washington, DC: Department of Transportation.

US Federal Aviation Administration (FAA). (2007). "Guide for Obtaining a Supplemental Type Certificate." Advisory Circular No. 21–40A. September 27. Washington, DC: Department of Transportation.

US Federal Aviation Administration (FAA). (2008a). "Assessment of FAA's Risk-Based System for Overseeing Aircraft Manufacturers' Suppliers." No. AV-2008-026. Washington, DC: Department of Transportation

US Federal Aviation Administration (FAA). (2008b). "Passenger Cabin Smoke Protection." Advisory Circular AC 25.795-4. Washington, DC: Department of Transportation.

US Federal Aviation Administration (FAA). (2009). "Bird Ingestion Certification Standards." Advisory Circular AC No: 33.76-1A. Washington, DC: Department of Transportation.

US Federal Aviation Administration (FAA). (2014). "Wildlife Strikes to Civil Aircraft in the United States 1990–2013." Federal Aviation Administration National Wildlife

Strike Database Serial Report Number 20. July. Washington, DC: Department of Transportation.

US Federal Aviation Administration (FAA). (2016). "A Study into the Structural Factors Influencing the Survivability of Occupants in Airplane Accidents." DOT/FAA/TC-16/31. Washington, DC: Department of Transportation.

US Fish and Wildlife Service (FWS). (2013). "Waterfowl Population Status." Washington, DC: US Department of the Interior.

US Government Accountability Office (GAO). (August 1992). "Aircraft Certification: Limited Progress on Developing International Design Standards." Report to the Chairman, Subcommittee on Aviation, Committee on Public Works and Transportation, Report No. 147597.

US Government Accountability Office (GAO). (1993). "Aircraft Certification: New FAA Approach Needed to Meet Challenges of Advanced Technology." GAO Report to the Chairman, Subcommittee on Aviation, Committee on Public Works and Transportation, House of Representatives. GAO/RCED-93-155, September 16.

US Government Accountability Office (GAO). (1995). "Aircraft Requirements: Air Force and Navy Need to Establish Realistic Criteria for Backup Aircraft." GAO/NSIAD-95-180. Report to Congressional Requesters. September.

US Government Accountability Office (GAO). (2004). "Aviation Safety: FAA Needs to Strengthen the Management of Its Designee Programs." Report to the Ranking Democratic Member, Subcommittee on Aviation, Committee on Transportation and Infrastructure. House of Representatives. October. http://www.gao.gov/cgi-bin/getrpt?GAO-05-40 (accessed March 10, 2023).

US Nuclear Regulatory Commission (NRC). (1983). "Handbook of Human Reliability Analysis with Emphasis on Nuclear Power Plant Applications." NUREG/CR—1278. Washington, DC: US Department of Energy.

US Nuclear Regulatory Commission (NRC). (2004). "Effective Risk Communication: The Nuclear Regulatory Commission's Guidelines for External Risk Communication." Report NUREG/BR-0308. January. Washington, DC: US Department of Energy.

US Nuclear Regulatory Commission (NRC). (2009). "Frequently Asked Questions about License Applications for New Nuclear Power Reactors." NUREG/BR-0468. Rockville, MD: Office of New Reactors, Nuclear Regulatory Commission.

US Nuclear Regulatory Commission (NRC). (2010). "Generic Issue 199 (GI-199): Implications of Updated Probabilisitic Seismic Hazard Estimates in Central and Eastern United States on Existing Plants." August. Washington, DC: US Department of Energy.

Van Maanen, J., and Pentland, B. (1994). "Cops and Auditors: The Rhetoric of Records." In Sitkin, S. B. and Bies, R. J. (Eds.), *The Legalistic Organization*. Newbury Park, CA: SAGE.

Vasigh, B., Flemming, K., and Humphreys, B. (2015). *Foundations of Airline Finance: Methodology and Practice*, 2nd ed. London: Routledge.

Vaughan, D. (1996). *The Challenger Launch Decision*. Chicago: University of Chicago Press.

Vaughan, D. (1999). "The Dark Side of Organizations: Mistake, Misconduct, and Disaster." *Annual Review of Sociology* 25: 271–305.

Vaughan, D. (2004). "Theorizing Disaster: Analogy, Historical Ethnography, and the Challenger Accident." *Ethnography* 5(3): 313–345.

Vaughan, D. (2005). "Organizational Rituals of Risk and Error." In Hutter, B. and Power, M. (Eds.), *Organizational Encounters with Risk*. New York and Cambridge: Cambridge University Press.

Vaughan, D. (2021). *Dead Reckoning: Air Traffic Control, System Effects, and Risk*. Chicago: University of Chicago Press.

Verran, H. (2012). "Number." In Celia, L. and Wakeford, N. (Eds.), *Inventive Methods: The Happening of the Social*. London: Routledge.

Villemeur, A. (1991). *Reliability, Availability, Maintainability and Safety Assessment*. Vol. 1 Chichester, UK: John Wiley & Sons.

Vincenti, W. G. (1979). "The Air Propellor Tests of W. F. Durand and E. P. Lesley: A Case Study in Technological Methodology." *Technology and Culture* 20: 712–751.

Vincenti, W. G. (1990). *What Engineers Know and How They Know It: Analytical Studies from Aeronautical History*. Baltimore: Johns Hopkins University Press.

Vincenti, W. G. (1994). "The Retractable Airplane Landing Gear and the Northrop 'Anomaly': Variation-Selection and the Shaping of Technology." *Technology and Culture* 35: 1–33.

Vincenti, W. G. (1997). "Engineering Theory in the Making: Aerodynamic Calculation 'Breaks the Sound Barrier." *Technology and Culture* 38: 819–825.

von Neumann, J. (1956). "Probabilistic Logics and Synthesis of Reliable Organisms from Unreliable Components." *Annals of Mathematics Studies* 34: 43–98.

Vosteen, L, and Hadcock, R. (1994). *Composite Chronicles: A Study of Lessons Learned in the Development, Production, and Service of Composite Structures*. Hampton, VA: National Aeronautics and Space Administration.

Waddington, T. (2000). *McDonnell Douglas DC-10*. Miami: World Transport Press.

Wall, M. (2016). "How a 1967 Solar Storm Nearly Led to Nuclear War." Space.com, August 9. https://www.space.com/33687-solar-storm-cold-war-false-alarm.html (accessed September 9, 2021).

Waltz, M. (2006). "The Dream of Composites." *R&D Magazine*, November 20. https://www.rdmag.com/article/2006/11/dream-composites (accessed April 4, 2019).

Wanhill, R. (2002). *Milestone Case Histories in Aircraft Structural Integrity*. NLR-TP-2002-521. Amsterdam: Nationaal Lucht- en Ruimtevaartlaboratorium.

Warwick, G. (1986). "Beech's Enterprising Starship." *Flight International*, May 3: 18–22.

WashingtonsBlog. (2013). "Fake Science Alert: Fukushima Radiation Can't Be Compared to Bananas or X-Rays." April 1. http://www.washingtonsblog.com/2013/04/fake-science-alert-fukushima-radiation-cant-be-compared-to-bananas-or-x-rays.html (accessed October 20, 2015).

Weart, S. (1988). *Nuclear Fear: A History of Images*. Cambridge, MA: Harvard University Press.

Weick, K. E. (1998). "Foresights of Failure: An Appreciation of Barry Turner." *Journal of Contingencies and Crisis Management* 6(2): 72–75.

Weick, K. E., and Sutcliffe, K. M. (2001). *Managing the Unexpected: Assuring High Performance in an Age of Complexity*. Jossey-Bass: San Francisco.

Weir, A. (2000). *The Tombstone Imperative: The Truth about Air Safety*. London: Simon & Schuster.

Wellock, T. R. (2017). "A Figure of Merit: Quantifying the Probability of a Nuclear Reactor Accident." *Technology and Culture; Baltimore* 58(3): 678–721.

Wellock, T. (2021). *Safe Enough? A History of Nuclear Power and Accident Risk*. Berkeley: University of California Press.

White, R. (2016). *Into the Black: The Electrifying True Story of How the First Flight of the Space Shuttle Nearly Ended in Disaster*. London: Transworld.

Wiener, E., Kanki, B., and Helmreich R. (Eds.). (1993). *Cockpit Resource Management*. San Diego: Academic Press.

Wildavsky, A. (1988). *Searching for Safety*. Oxford, UK: Transaction.

Wiley, J. (1986). "A Capture Theory of Antitrust Federalism." *Harvard Law Review* (99): 713–723.

Wilson, A. (1973). *The Concorde Fiasco*. London: Penguin Special.

Wittgenstein, L. (2001 [1953]). *Philosophical Investigations*. London: Blackwell Publishing.

Wolf, F. (2001). "Operationalizing and Testing Normal Accident Theory in Petrochemical Plants and Refineries." *Production and Operations Management* 10: 292–305.

World Nuclear Association. (2019). "Nuclear Power in the World Today." https://www.world-nuclear.org/information-library/current-and-future-generation/nuclear-power-in-the-world-today.aspx (accessed August 6, 2019).

Wright, T.P. (1936). "Factors Affecting the Cost of Airplanes." *Journal of the Aeronautical Sciences* 3: 122–128.

Wu, J. S., and Apostolakis, G. E. (1992). Experience with Probabilistic Risk Assessment in the Nuclear Power Industry. *Journal of Hazardous Materials* 29(3): 313–345.

Wynne, B. (1988). "Unruly Technology: Practical Rules, Impractical Discourses and Public Understanding." *Social Studies of Science* 18: 147–167.

Wynne, B. (1989). "Frameworks of Rationality in Risk Management: Towards the Testing of Naive Sociology." *Environmental Threats: Social Sciences Approaches to Public Risk Perceptions*: 33–45.

Wynne, B. (2003). "Seasick on the Third Wave? Subverting the Hegemony of Propositionalism." *Social Studies of Science* 33(3): 401–417.

Wyss, G. (2016). "The Accident That Could Never Happen: Deluded by a Design Basis." In Sagan, S. and Blandford, E. (Eds.), *Learning from a Disaster*. Stanford, CA: Stanford University Press.

Yanagisawa, K., Imamura, F., Sakakiyama, T., Annaka T., Takeda, T., and Shuto, N. (2007). "Tsunami Assessment for Risk Management at Nuclear Power Facilities in Japan." *Pure and Applied Geophysics* 164: 565–576.

Younossi, O., Kennedy, M., Graser, J.C. (2001). *Military Airframe Costs: The Effects of Advanced Materials and Manufacturing Processes*. RAND Corporation, Santa Monica, CA.

Zdzislaw, K, Szczepanski, P., and Balazinski, M. (2007). "Causes and Effects of Cascading Failures in Aircraft Systems." *Diagnostyka* 1(41): 19–26.

Zimmel, T. (March 2004). "Quality Science: A Historical Perspective. Part 1: The Early Years." http://www.msi.ms/MSJ/QUALITY_historical_1_20000603.htm (accessed July 5, 2014).

INDEX

Inside Technology Series

Edited by Wiebe E. Bijker and Rebecca Slayton

Helga Nowotny, *Insatiable Curiosity: Innovation in a Fragile Future*

Karin Bijsterveld, *Mechanical Sound: Technology, Culture, and Public Problems of Noise in the Twentieth Century*

Peter D. Norton, *Fighting Traffic: The Dawn of the Motor Age in the American City*

Joshua M. Greenberg, *From Betamax to Blockbuster: Video Stores and the Invention of Movies on Video*

Mikael Hård and Thomas J. Misa, editors, *Urban Machinery: Inside Modern European Cities*

Christine Hine, *Systematics as Cyberscience: Computers, Change, and Continuity in Science*

Wesley Shrum, Joel Genuth, and Ivan Chompalov, *Structures of Scientific Collaboration*

Shobita Parthasarathy, *Building Genetic Medicine: Breast Cancer, Technology, and the Comparative Politics of Health Care*

Kristen Haring, *Ham Radio's Technical Culture*

Atsushi Akera, *Calculating a Natural World: Scientists, Engineers and Computers during the Rise of U.S. Cold War Research*

Donald MacKenzie, *An Engine, Not a Camera: How Financial Models Shape Markets*

Geoffrey C. Bowker, *Memory Practices in the Sciences*

Christophe Lécuyer, *Making Silicon Valley: Innovation and the Growth of High Tech, 1930–1970*

Anique Hommels, *Unbuilding Cities: Obduracy in Urban Sociotechnical Change*

David Kaiser, editor, *Pedagogy and the Practice of Science: Historical and Contemporary Perspectives*

Charis Thompson, *Making Parents: The Ontological Choreography of Reproductive Technology*

Pablo J. Boczkowski, *Digitizing the News: Innovation in Online Newspapers*

Dominique Vinck, editor, *Everyday Engineering: An Ethnography of Design and Innovation*

Nelly Oudshoorn and Trevor Pinch, editors, *How Users Matter: The Co-Construction of Users and Technology*

Peter Keating and Alberto Cambrosio, *Biomedical Platforms: Realigning the Normal and the Pathological in Late-Twentieth-Century Medicine*

Paul Rosen, *Framing Production: Technology, Culture, and Change in the British Bicycle Industry*

Maggie Mort, *Building the Trident Network: A Study of the Enrollment of People, Knowledge, and Machines*

Donald MacKenzie, *Mechanizing Proof: Computing, Risk, and Trust*

Geoffrey C. Bowker and Susan Leigh Star, *Sorting Things Out: Classification and Its Consequences*

Charles Bazerman, *The Languages of Edison's Light*

Janet Abbate, *Inventing the Internet*

Herbert Gottweis, *Governing Molecules: The Discursive Politics of Genetic Engineering in Europe and the United States*